Smart Society

Increasingly, we hear of 'smart' cities, communities, governance and people as constituting the basis of initiatives by which we might address various social and environmental problems, particularly those connected with sustainability, usually by means of an 'intelligent' connection with the 'network society'. This book addresses the issues raised by the emergence of 'smart' dimensions and initiatives in society, critically engaging with questions surrounding the feasibility of what smart initiatives propose and the extent to which they can really offer solutions to the challenges we face. With attention to the notion of 'smart' as applied to the individual, the community, politics and the home, the authors consider the interconnections between these various facets of 'smart living' and their relationship to the notion of the smart society as a whole. Drawing on a concrete study of an attempt to concretize smart ideas in the design of a smart, solar home as part of an international project, *Smart Society* offers the first extended sociological engagement with the notion of smart living.

Roberta Iannone is Associate Professor of Sociology in the Department of Political Science at the Sapienza University of Rome, Italy. She is author of *The social sense of the human experience: Thinking about Vom Menschen of Werner Sombart*, and *Il capitale sociale. Origine, significati e funzioni* (2006).

Romina Gurashi is Subject Expert in Sociology at the Sapienza University of Rome, Italy. She is Managing Editor of the *Quarterly Journal of Administration Science. Studies in Theory and Social Research*, the author of *Pathways of Peace: Philosophy and Sociology of Peace and Nonviolence* (2018) and the co-author of *From Intractability to Pacification: A Federal perspective for the Abkhaz-Georgian Conflict* (2017).

Ilaria Iannuzzi is completing her doctoral studies in the Department of Political Sciences at the Sapienza University of Rome, Italy. She is the author of *La fiducia paga. Quando le relazioni generano valore* (*Trust pays. When relationships generate value*) and of several articles in scientific journals. She is an accredited Professional by Italy-USA Foundation.

Giovanni de Ghantuz Cubbe is Ph.D Student and Research Associate at the Institute of Political Science, TU Dresden, Germany. His research interests focus on contemporary nationalism, migration and European radical right parties as well as on structural and conceptual changes of democracy and political institutions in Europe.

Melissa Sessa is the best-graduated student of July 2017 in Administration and Organization Sciences. She is currently completing a master's degree in Political Science. She is the winner of the prestigious prize for a dissertation thesis, "Vincenzo Dona", with her work on *La smart home nella sua dimensione sociale* (*The smart home in its social dimension*).

Smart Society

A Sociological Perspective on
Smart Living

**Roberta Iannone, Romina Gurashi,
with Ilaria Iannuzzi,
Giovanni de Ghantuz Cubbe and
Melissa Sessa**

Routledge
Taylor & Francis Group

LONDON AND NEW YORK

First published 2020
by Routledge
2 Park Square, Milton Park, Abingdon, Oxon OX14 4RN

and by Routledge
52 Vanderbilt Avenue, New York, NY 10017

Routledge is an imprint of the Taylor & Francis Group, an informa business

British Library Cataloguing in Publication Data
A catalogue record for this book is available from the British Library

Library of Congress Cataloging-in-Publication Data
A catalog record has been requested for this book

ISBN: 978-0-367-19241-9 (hbk)
ISBN: 978-0-429-20127-1 (ebk)

Typeset in Times New Roman
by Taylor & Francis Books

Contents

1 Smart society

The critical sense of a world strategy

Roberta Iannone

Around and within the 'smart' world

By virtue of a certain information redundancy, the adjective 'smart' is now monopolizing the scientific literature in every disciplinary field: from engineering to economics and political science, from sociology to urban planning. Even the pages of the newspapers use the term 'smart' at least as frequently as the academic congresses. The conversations between scientists as well as between journalists, between politicians and between citizens, between consumers and between producers find in this adjective one of the most used and abused words. The translations of the term are often varied, its interpretations disparate and its meanings polysemic.

The city is certainly the most natural habitat of the smart world and the literature on smartness is dominated by the 'smart city' (Aa.Vv. 2014; Marciano 2015), to the point that almost all theoretical insights focus on this meaning (De Luca 2012) and on the relative distinctions. This is the case of the city of networks (or net city), of the open city, of the sentient city, of the participatory city (or wiki city), of the neo-bohemian city (or creative city) and, again, the resilient city, the 2.0 city, and finally the city as a platform (or cloud city) (Dominici 2012).

Beyond or next to the city, according to the most consolidated literature on the subject (Amitrano and Bifulco 2016), it is possible to trace at least six declinations of 'smartness':

1 smart economy
2 smart people
3 smart governance
4 smart mobility
5 smart environment
6 smart living

In economics (Balaceanu et al. 2017), the word smart refers to more dynamic and competitive, innovative and entrepreneurial activities.

Smart people (Barrett 2017; Bates and Gupta 2017; Gurashi 2018; Kar et al. 2017) are instead the expression of citizens who are creative and flexible, but also professionally qualified and open to forms of participation and social integration.

The political participation of citizens in the decision-making process is instead the distinctive element of 'smart governance' (Vinod Kumar 2015; Willke 2007), where the participation in the creation of public services contributes to making governance more transparent and democratic.

The 'smart mobility' (European Environment Agency 2016) refers, instead, to the transport system, to the national and international accessibility of mobility and to the possibility of making it innovative and safe.

Finally, the 'smart environment' (European Environment Agency 2017; Urban Land Institute 1998) refers to a sustainable management of natural resources and to the preservation and protection of the green areas of a city. 'Smart Community' (D'Aloisi et al. 2013; Manfredi 2015; Rizzi 2014; Urban Land Institute 1998), 'smart land' (Bonomi and Masiero 2014), 'smart home' (Balta-Ozkan et al. 2013; Briere 2003; Capolla 2011; Gurashi et al. 2019; Harper 2003; Saul-Rinaldi et al. 2014; Sessa 2018) and 'smart working' (Corso et al. 2016; Lake 2013) complete, then, the framework of the possible variations of the concept. But what is the common denominator of all these 'smart' worlds?

Thinking smart is first of all thinking in an 'integrated' way and the main integration is that which is realized between people, environment and technologies. We could long argue about the centrality of information technology with respect to smartness because indeed without technology none of these forms of intelligence would be possible. And yet the smart city, and with it all its variations of smartness, is not a mere digital city. 'Smart' and 'digital' are not synonyms, but rather they allude to diversified even intersected plans and the smart city is something more than a mere digital city (Aurugi 2005; Dameri and Giovannacci 2015) or information city (Mola et al. 2015).

In this sense, the 'smart society' (Hayman 1998; Mallapaty 2018; Ramachander 2018; Valkenburg et al. 2016; Voronkova and Kyvliuk 2017) becomes central. The smart society is the real protagonist and it is only within a systemic logic that it is possible to think and act 'smart'. Society is the social system that makes every smart initiative possible, whether it is economic, political, working or housing; plus, it integrates people, environment and technologies. It is that amalgam, that cohesion, that set of ties and relationships, that intersection of plans and social formations that only makes the smart dimension possible, beyond any possible declination.

Understanding 'smartness' therefore means understanding the dimension of integration that comes before any form of 'intelligence', making it possible. A new systemic integration based on digital technologies and aimed at saving energy.

(Gabrielli and Granelli 2014)

Around and within the smart world there is then a very strong and constitutive reference to the idea of *intelligent growth* that is, at the same time, also *sustainable growth* and *inclusive growth*. This is what emerges from the 2020 Horizon program (the European Union Framework Program for Research and Innovation) and from the 2030 Agenda for sustainable development (the Global Program for Sustainable Development) which constitute the referential regulatory and value frameworks in this regard.

From theory to practice: the heart of the question

However, if we move from theory to practice, the question becomes more complicated and the difficulties seem to lie precisely in a joint declination of these elements. The impression that moves these pages is that often the goal of smartness understood as 'intelligent growth' – no matter what the level on which it stands (micro, meso, macro) or the field of experience with which one desires to decline it (home, city, work, etc.) – is separated from sustainable growth and inclusive growth. To say, through the absolutizations of the economy, albeit based on knowledge and innovation, uprooting it from a greener economy, as well as from an economy capable of employment and cohesion. In this way, not only the smartness becomes synonymous with mere technological efficiency, but it loses all the 'intelligence' charge that it had promised. At that point it will not appear unfounded to wonder why using a new term, such as smart precisely, is better than instead of using the more traditional terms of technological, digital or simply efficient society.

It is therefore opportune to rest and reflect on the somewhat 'holistic' character that the term 'smart' brings with respect to the triple dimension of sustainability on the one hand and of inclusiveness on the other. The word 'smart' itself, if combined with intelligence, can lead to a truly integrated, smart society.

The impression is that, beyond the often very fascinating rhetoric of the smart world as a technological and competitive world, nothing can ever be truly 'intelligent' if technology and economy do not find their place in society. It is the place with which both these dimensions of social experience were born and which for millennia has given them a function for the guarantees of order and development of society.

This aspect, debated but still minor in the dominant literature and which needs to be brought to light in terms of sociological theory and empirical research attentive to the ensuing practical implications, is central to every realistic idea of smart society. It is not possible to talk about the smart society if we are not confronted with the problem of the current and potential boundaries of technology and economy. And if these boundaries continue to be read with the technical or engineering look, and not in reference to that system of needs and interests that we call society and to that axiological cultural heritage that we call culture.

It is a question of rethinking the cultural and social sense that economy and technology have not only with respect to the needs of order, but also for progress (more and better than growth). An engineering use of new technologies is not enough, because efficiency is not in itself a guarantee of development. To be intelligent, the use of technology must be based on a vision upstream of the relationship between technology and society, which re-establishes the original terms of the relationship and the placement of the techné in society. Technique, like technology, has always been considered an indispensable condition for human existence. On the contrary: they can be considered as the very essence of man (Galimberti 2009). And this is because man, not being codified by instincts like animals, can survive only if he immediately becomes technician. When, however, the technique from medium becomes an end in itself, it is likely to assume that one enters the realm of technicalities as a method to transform a technique into a given abstract and general system (Postman 1981).

The same can be said for economy. Born as a sphere between the spheres aimed at satisfying the needs of the community and its possibilities of development, economy currently finds itself as *the* dominant sphere. It emerges from its original location, which saw it as part of a larger social system, or more precisely a social subsystem. When this happens, the market imposes itself and from an institution immersed into the social sphere, it becomes the institution that forms society and expresses a certain economic hegemony. The market becomes the scene of representation of the whole sociability, it becomes the formula through which the quantitative objectification of exchange is realized in modern culture. Likewise, among the key values of late modernity, profit and efficiency are affirmed far beyond economic action, invading every area of social experience, and invalidating the economic value itself, which is increasingly self-referential, rather than constructed and oriented towards society. The economy then becomes economism, a real ideology that informs every area of associated life (Mongardini 1997).

As Pareto has already noted, and as we should never forget, there is the profit *of* society but also the profit *for* society: the former is not always compatible with the latter and we must be careful not to confuse "the maximum utility *for* a collectivity, with the maximum utility *of* a community" (1916: par. 2133). In the first case, it is the utilities of the individual that prevail. In the second case it is society that is considered as a unit whose purpose prevails over individual ends.

This is the framework within which it is considered appropriate to frame the theme of the smart society and its actual concretizations. A framework that is broader than the one that follows individual acts or specific behaviors in the technical and economic sphere, without prejudice to the centrality of technique/technology and the economy for the smart dimension to succeed in its consolidation.

Hence, therefore, the conceptual continuum that this book intends to investigate. It considers the problems of the smart society mostly linked to the distortions inherent in the sense of the economy, in the direction of economics and the economy 'as an ideology', and of technology in the sense given by technicalization and technicality. Consequently, the smart society becomes economy and technique more 'aware' of their role, with related limits and potential, through a more conscious economy and a more intelligent technique/technology.

We still need to understand which 'smart' areas must be perfected for this to happen *de facto* and beyond the declarations of principle, creating an integrated social order.

Conclusions

In the performative society that smartness brings with it, production efficiency is essential. But is efficiency possible without integration?[1] If we want to verify the ideal of efficiency, we must focus on integration. And on an integration that is not mere agreement between the interests of a stakeholder society, but true social cement. A mechanical and organic solidarity together, Durkheim would say. The threats, however, are known and go under the names of hyperspecialism and consequent social fragmentation, solipsism or individual atomization. Added to this are the threats of a connection that is not real integration and proper to a hyperconnected world and a perpetually 'linked' or hypertextual society, in which everything that connects does not penetrate by definition. It is precisely this integration that is affected, which instead is the extreme synthesis of concepts as extremely interconnected and interdependent phenomena.

The concept of 'integration' must therefore be recovered and given back to the literature on smartness in all its undoubted centrality. With this concept the question that contemporary societies are faced with will appear even clearer: can organicistic logic, which divides and combines, which makes the parts specialized but also interdependent, which binds through complementary interest, be enough to guarantee unity, or we need even more, because interdependence and mutual interest alone are not enough, especially if it is an interest with no shared identity and limited to the content of efficiency alone. It is no longer time, on the other hand, to choose between universalism on the one hand and particularism on the other. Perhaps the time has come to realize "the universalism of attribution" and the "particularism of realization" (Parsons 1965), as only a joint, and smart, declination of community and society can do. Therefore a smart community, properly understood, becomes central, as we will try to describe in the following pages (Iannuzzi 2019).

Giving back to the concept of integration the weight and centrality that it deserves, it is also possible to avoid that the smartness becomes a mere synonym of a *performing society in the era of records* and that smart people take on the role of children (or mere consumers) who find all their identity[2] in 'material greatness', 'rapidity of movement', 'perennial novelty' and 'sense of power' (Sombart 1978: 136). It is no coincidence that specific attention will be devoted in this book to smart people and their role for a more promising smart society (Gurashi 2019).

Efficient hyperbole would make smartness, not only short-sighted but also harmful because it is capable of 'perverse effects'. New technologies can in fact allow not only the fight against social exclusion, but they can also be a source, creating new forms of digital divide and lack of access to services. In this case, would we still be faced with a 'smart' reality? If it is not accompanied by measures that make the innovation accessible and usable, the latest discoveries of the technology applied to the experience are not in themselves a guarantee of improvement or development, as exemplified by the case of the smart homes (Sessa 2019). Likewise, a technology that facilitates certain operations, but with high social and economic costs, cannot be called a smart technology, but simply an efficient technology.

These are, perhaps, the reasons that have led to the need to go beyond digitalization, in theory as in practice, through the smart perspective. However, no cognitive advantage can derive from the use of this construct if it remains undetermined and confused. The risk of making smartness merely synonymous with technological and/or digital advancement always lies in wait and this should push towards a contextual advancement both of material and non-material culture (Ogburn 1922), even if only to not leave important critical issues in the background. This is the case referred to by Hollands (2008; De Luca 2012) of the implications of the entrepreneurial city described by Harvey (1989), of the domination of the neoliberal spaces denounced by Peck and Tickell (2002), of discrimination between more ordinary citizens and more citizens 'performance' described by Amin, Massey and Thrift (2000), or problems related to urban marketing according to Short (Short et al. 2000).

The 'joint optimization' of the technical and social functions of a system then becomes central and an objective to avoid that the strengthening of technology involves an underestimation of the social sphere. This is the case with the optimization of the technical-economic performance that can easily lead to poor results on other fronts with undoubted repercussions on what was intended to be strengthened. In fact, often ignoring or underestimating something means making it re-emerge in a radicalized and extreme form. Even at the cost of slower steps or goals (at least in the immediate future), the joint optimization of the functions is therefore the aim towards which a full and promising smartness should strive to fully realize itself. Exactly as it happens in the field of natural resources where the '(sub)systemic lesson' seems to be acquired in terms of smart environment.

The smart society is a sustainable society as it is integrated in all its levels, in its structures and areas of experience. If it is true that smart:

> …stands for efficient, capable, inclusive, modern, sustainable, therefore starting from the original footprint of ICT infrastructures, an intelligent city [like a smart society] must also include coordinated and integrated interventions at a social, environmental and economic level aimed at enhancing human capital, reducing environmental impacts and resolving environmental emergencies considered to be priorities (for example land consumption, urban and energy redevelopment, mobility, waste management) with related economic benefits.
>
> (Federico 2013: 35)

Therefore, the question that perhaps only 'the general system theory' has been able to address – at least in its problematic nature and regardless of the gradually changing answers – is the assumption according to which no system analysis (and we should speak of system analysis in the case of smartness) can be carried out unless it is specified by whom and how the unit of government is constituted, what are the aims towards which it tends and which of these aims are consolidated by certain kinds of forms of control and social regulation (Luhmann 1984). That is, no system analysis is possible if the ideal of 'transparent society' is not proven true (Vattimo 2000). The hyper-technological guise of postmodern society in which the media plays a decisive role in making the experience more chaotic and fostering new hopes of emancipation is not sufficient in itself. "Today humanity must rise to the level of its technological possibilities", says Vattimo (2000), but also "imagine an ideal of man" (Vattimo 2000) – and at this point we could also add 'relations' – "that takes account and fully utilize these possibilities". In order for this to happen, as we will have the opportunity to grasp in the following pages (de Ghantuz Cubbe 2019), we need a management of the legitimate power, and more generally of politics and 'governance', that is equally smart and up to the task, i.e. that is "smart, transparent, inclusive, capable of developing a clear and shared vision of well-being, quality of life and sustainability" (Federico 2013: 39) and above all capable of being democratic. No smart society is therefore possible without a smart politics, conscious and attentive to power relations (always present within the algorithms) and to the politicity that is inherent in the smart society, even when, or perhaps especially when it is not institutionalized. A smart society capable of government models based on the centrality of relational and common goods and on civic participation in the creation of value, and able to be simultaneously a smart community, in which connection and sharing, i.e. active relational connections enhanced by the use of new technologies, and a dynamic and adaptive production of common sense, oriented to participation, constitute innovative forms of connective empowerment (Paini 2012). The following pages are dedicated to these scenarios and the empirical case seems to be a proof of its practicability.

Notes

1 In this regard, L. Bruni writes: "The economy of a country depends above all on its capital. During the second half of the twentieth century, Italy was capable of a real economic and civil miracle because it had social, moral, spiritual, community capitals, and the system as a whole was capable of generating income. We would not have transformed a country with widespread poverty into one of the world's economic powers without those assets (the gift of the fathers: *patres munus*) made of civil virtues, of the value of sacrifices, of faith, of ideals; in the 1970s we would not have tripled the number of companies from 300 thousand to one million without the peasant and artisan ethics of well-done work. Without forgetting that immense capital made of and caring for women: an enormous heritage, unrecognized and unpaid" (Bruni 2016). On these themes, see the essay by Iannuzzi (2018).

2 "I do not think it is unlikely", Sombart warned at the beginning of the 1900s, "that in a few hundred years the historian who will have to describe the time in which we live today, puts the title at the head of that part of his book: the age of records" (1978: 136).

References

Aa.Vv. (2014) "Smart city. Città, tecnologia, comunicazione." *Comunicazione-puntodoc* 10, 1–284.

Amin, A., Massey, D. and Thrift, N. (2000) *Cities for the many not for the few*. Bristol: Policy Press.

Amitrano, C. C. and Bifulco, F. (2016) "Level of smartness in urban contexts: open issues in measurement." 26th Annual RESER Conference, What's Ahead in Service Research? New perspectives for business and society, Naples, 590–603.

Aurugi, A. (2005) *Making the digital city: the early shaping of urban internet space*. Burlington: Ashgate.

Balaceanu, C., Tilea, D. M. and Penu, D. (2017) "Perspectives on Eco Economics. Circular Economy and Smart Economy." *Academic Journal of Economic Studies* 3(4), 105–109.

Balta-Ozkan, N., Davidson, R., Bicket, M. and Whitmarsh, L. (2013) "Social barriers to the adoption of smart homes." *Energy Policy* 63, 363–374. doi:10.1016/j.enpol.2013.08.043.

Barrett, C. (2017) "Smart People, smart ideas and the right environment drive innovation." *Research Technology Management* 53(1), 40–43. doi:10.1080/08956308.2010.11657610.

Bates, T. C. and Gupta, S. (2017) "Smart groups of smart people: Evidence for IQ as the origin of collective intelligence in the performance of human groups." *Intelligence* 60, 46–56.

Boccia Artieri, G. (2012) *Stati di connessione: pubblici, cittadini e consumatori nella (social) network society*. Milano: FrancoAngeli.

Bonomi, A. and Masiero, R. (2014) *Dalla smart city alla smart land*. Venezia: Marsilio.

Briere, D. D. (2003) *Smart home for dummies*. New York: Frommer's.

Bruni, L. (2016) "Investire su humanities e coesione sociale." *Il Sole 24 Ore*, 2 March.

Capolla, M. (2011) *Progettare la domotica. Criteri e tecniche per la progettazione della casa intelligente*. Santarcangelo di Romagna: Maggioli Editore.

Corso, M., Crespi, F. and Sacco, A. C. (2016) *Smart working: modelli organizzativi e tecnologie, spazi e normativa*. Milano: Gruppo 24 ore.

D'Aloisi, D., Persia, S. and Sapio, B. (2013) "Smart Community: l'evoluzione sociale della Smart City." *I quaderni di Telèma*. http://www.fub.it/sites/defa ult/files/attachments/2013/09/n295.pdf (January 12, 2019).

Dameri, R. P. and Giovannacci, L. (2015) *Smart city e digital city: strategie urbane a confronto*. Milano: FrancoAngeli.

de Ghantuz Cubbe, G. (2019) "Smart politics. The political dimension of the 'smartness'." In Iannone, R. and Gurashi, R. with Iannuzzi, I., de Ghantuz Cubbe, G. and Sessa, M., *Smart Society. A Sociological Perspective on Smart Living*. Abingdon: Routledge.

De Luca, A. (2012) "Come (ri)pensare la smart city." *EyesReg. Giornale di Scienze Regionali* 2(6), 143–146.

Dominici, G. (2012) "Smart cities e communities: l'innovazione nasce dal basso." http://archive.saperi.forumpa.it/ (January 12, 2019).

Duany, A., Lydon, M. and Speck, J. (2010) *The smart growth manual.* New York: McGraw-Hill.

Durkheim, É. (1893) *De la division du travail social.* Paris: Félix Alcan Editeur.

European Environment Agency (2016) *Towards clean and smart mobility: transportation and environment in Europe.* Luxembourg: Publications Office of the European Union.

European Environment Agency (2017) *Shaping the future of energy in Europe: clean, smart and renewable.* Luxembourg: Publications Office of the European Union.

Federico, T. (2013) "Smart city: innovazione e sostenibilità." *EAI. Energia, Ambiente e Innovazione* 5, 35–40.

Freilich, R. H. (1999) *From sprawl to smart growth: successful legal, planning, and environmental systems.* Chicago: American Bar Association.

Gabrielli, G. and Granelli, A. (eds.) (2014) *Territori, città, imprese: smart o accoglienti?* Milano: FrancoAngeli.

Galimberti, U. (2009) *I miti del nostro tempo.* Milano: Feltrinelli.

Gallino, L. (1978) *Dizionario di sociologia.* Torino: UTET.

Gil-Garcia, J. R., Zhang, J. and Puron-Cid, G. (2016) "Conceptualizing smartness in government: an integrative and multi-dimensional view." *Government Information Quarterly* 33(3), 524–534. doi:10.1016/j.giq.2016.03.002.

Gurashi, R. (2018) "The era of the smart people. How technocapitalism is changing the lifestyles of the individuals of the smart society." *Storiadelmondo* 86, 1–12.

Gurashi, R. (2019) "Smart people: the individual challenge of the fourth industrial revolution." In Iannone, R. and Gurashi, R. with Iannuzzi, I., de Ghantuz Cubbe, G. and Sessa, M., *Smart Society. A Sociological Perspective on Smart Living.* Abingdon: Routledge.

Gurashi, R., Iannuzzi, I. and Sessa M. (2019) "From theory to empiricism: the challenge to the future of the Sapienza University at Solar Decathlon Middle East." In Iannone, R. and Gurashi, R. with Iannuzzi, I., de Ghantuz Cubbe, G. and Sessa, M., *Smart Society. A Sociological Perspective on Smart Living.* Abingdon: Routledge.

Harper, R. (2003) *Inside the Smart Home: Ideas, Possibilities and Methods.* London Ltd: Springer-Verlag.

Harvey, D. (1989) "From managerialism to entrepreneurialism: the transformation in urban governance in late capitalism." *Geografiska Annaler* 71b(1), 3–17. doi:10.1080/04353684.1989.11879583.

Hayman, R. (1998) *The smart culture: society, intelligence, and Law.* New York: New York University Press.

Hollands, R. G. (2008) "Will the real smart city please stand up?" *City* 12(3), 303–320. doi:10.1080/13604810802479126.

Iannuzzi, I. (2018) "The smart city. Critical reading of a multiform phenomenon." *Storiadelmondo* 86, 1–13.

Iannuzzi, I. (2019) "Smart community: a new way of being together?" In Iannone, R. and Gurashi, R. with Iannuzzi, I., de Ghantuz Cubbe, G. and Sessa, M., *Smart Society. A Sociological Perspective on Smart Living.* Abingdon: Routledge.

Kar, A. K., Gupta, M. P., Ilavarasan, P. V. and Dwivedi, Y. K. (2017) *Advances in Smart Cities: Smarter People, Governance, and Solutions.* Boca Raton: CRC Press.

Lake, A. (2013) *Smart flexibility: moving smart and flexible working from theory to practice.* Gower, Farnham: Burlington.

Luhmann, N. (1984) *Soziale Systeme: Grundriss einer allgemeinen Theorie.* Frankfurt: Suhrkamp.

Mallapaty, S. (2018) "Pillars of a smart society." *Nature* 555(7697), S62–S63. doi:10.1038/d41586-41018-02899-x.

Manfredi, F. (2015) *Smart community: comunità sostenibili e resilienti.* Bari: Cacucci.

Marciano, C. (2015) *Smart City. Lo spazio sociale della convergenza.* Roma: Edizioni Nuova cultura.

Mezzapelle, D. (2016) "Smartness come 'stile di vita'. Approcci alla discussione, Bollettino della Società Geografica Italiana." *Roma* Serie XIII(IX), 489–501.

Moccia, F. (2012) "Smart city: etimologia del termine. Un'analisi firmata INU." http://admin.edilio.it/smartcity-etimologia-del-termine-un-analisi-firmata-inu/p_19560.html (January 12, 2019).

Mola, L., Pennarola, F. and Za, S. (2015) *From Information to Smart Society. Environment, Politics and Economics.* Berlin: Springer International Publishing.

Mongardini, C. (1997) *Economia come ideologia. Sul ruolo dell'economia nella cultura moderna.* Milano: FrancoAngeli.

Ogburn, W. F. (1922) *Social Change with Respect to Culture and Original Nature.* New York: Viking Press.

Paini, G. (2012) "Cosa sono le smart communities." http://www.thinktag.it/system/files/11778/Smart_Commu.pdf (accessed January 12, 2019).

Papa, R., Fistola, R. and GargiuloC. (eds.) (2018) *Smart planning: sustainability and mobility in the age of change.* Cham: Springer.

Papa, R., Gargiulo, C., Franco, S. and RussoL. (2014) "Urban smartness vs urban competitiveness. A comparison of Italian cities rankings." *TeMA. Journal of Land Use, Mobility and Environment* 7, 771–782.

Pareto, V. (2016) *Trattato di sociologia generale, vol. II.* Firenze: G. Barbera editore.

Parsons, T. (1965) *Il sistema sociale.* Milano: Edizioni di Comunità.

Peck, J., Tickell, A. (2002) "Neoliberalising space." *Antipode* 34(3), 380–404.

Postman, N. (1981) *Ecologia dei media.* Roma: Armando.

Ramachander, R. (2018) "Smart technology, smart society." https://www.powerengineeringint.com/articles/2018/07/in-depth-smart-technology-smart-society-big-data-in-energy.html (January 12, 2019).

Rizzi, F. (2014) *Smart city, smart community, smart specialization per il management della sostenibilità.* Milano: FrancoAngeli.

Rydén, L. (2015) "Technological Development and Lifestyle changes." In Leal Filho, W., Úbelis, A. and Bērziⓧa, D. (eds.), *Sustainable Development, Knowledge Society and Smart Future Manufacturing Technologies*. World Sustainability Series. Switzerland: Springer International Publishing, 113–124.

Saul-Rinaldi, K., Lebaron, R. and Caracino, J. (2014) *Making sense of the Smart Home. Applications of Smart Grid and Smart Home Technologies for the Home Performance Industry.* USA: National Home Performance Council.

Schiffman, I. (2001) *Alternative techniques for managing smart growth.* University of California, Berkeley: Berkeley Public Policy, Institute of governmental studies.

Sessa, M. (2018) "The social dimension of the smart home. How sustainability became part of the domestic environment." *Storiadelmondo* 86, 1–10.

Sessa, M. (2019) "Home smart home. A sociology of living in the age of reflective materialism." In Iannone, R. and Gurashi, R. with Iannuzzi, I., de Ghantuz Cubbe, G. and Sessa, M., *Smart Society. A Sociological Perspective on Smart Living*. Abingdon: Routledge.

Short, J. R., Breitbach, C., Buckman, S. and Essex, J. (2000) "From World Cities to Gateway Cities: Extending the Boundaries of Globalization Theory." *City* 4 (3), 317–340.

Sombart, W. (1978) *Il borghese. Contributo alla storia intellettuale e morale dell'uomo economico moderno*. Milano: Longanesi.

Szold, T. S. and Carbonell, A. (eds.) (2002) *Smart growth: form and consequences.* Cambridge (Mass.): Lincoln Institute of Land Policy.

Urban Land Institute (1998) *Smart growth: economy, community, environment.* Washington DC: Urban Land Institute.

Valkenburg, A. C., den Ouden, P. H. and Schreurs, M. A. (2016) *Designing a smart society: from smart cities to smart societies*. Netherlands: European Commission.

Vattimo, G. (2000) *La società trasparente.* Milano: Garzanti.

Vinod Kumar, T. M. (ed.) (2015) *E-governance for smart cities.* Singapore: Springer.

Vitari, C. (2015) "Social Equity and Ecological Sustainability: New Framework and Directions for the IS Community." In Mola, L., Pennarola, F. and Za, S., *From Information to Smart Society. Environment, Politics and Economics*. Berlin: Springer International Publishing.

Voronkova, V. and Kyvliuk, O. (2017) "Philosophical Reflection Smart-Society as a New Model of the Information Society and its Impact on the Education of the 21st Century." *Future Human Image* 7, 154–162.

Willke, H. (2007) *Smart governance: governing the global knowledge society.* Frankfurt: Campus Verlag.

2 Smart politics

The political dimension of 'smartness'

Giovanni de Ghantuz Cubbe

Introduction

The concept of politics, in addition to having been defined by its etymological root, indicating in general the management of public activity, has been the subject of multiple interpretations, whose sole outline would require a whole work. It has, indeed, an incredible variety of meanings, depending on the historical context in which it has been used (from ancient Greece, passing through Rome and up to modernity) and can generally indicate different organizational and institutional structures and forms of government (political systems), can be considered as a specific form of social action, or, in a more restricted sense, only as one 'dimension' of the political world (an example is the Anglo-Saxon tradition with its subdivision into polity, *politics*, policy). In this context, our aim is to observe politics first of all as a complex social phenomenon linked to the meaning and the form of power in a given social organization. In the moment in which social arrangements and interactions of power take shape and establish hierarchies to which an interpretation, a symbolic value, a form of legitimacy is attributed, or resistance is opposed to them, political life is already beginning. In this sense, politics is a dynamic intertwining of interactions, de facto structures and relations of power on the one hand and attributions of meaning on the other (Burdeau 1979; de Ghantuz Cubbe 2015; Mongardini 2011), be it the latter civil religions, traditional values, or 'sacralised' political ideologies (Cladis 2010; de Ghantuz Cubbe 2017; Miccoli 2010). Thus understood, politics is part of every social organization and every discourse around it. In this sense, politics is also part of the concept of smartness, since it represents (also) a discourse around social organization. However, the so-called *smart society* (or, better to say, the general, aggregate and systemic dimension of 'smartness' in a society (Iannone 2019)), has long been considered as 'post-political' (Wiig and Wyly 2016: 487) and was therefore often overlooked by political science.

Our will here is to support two theses:

1   Wherever there is a smart society – whether it is only conceived in theory or concretely developed in practice – there is also a 'smart politics', understood in general as the set of structures and dynamics of power and their symbolic interpretation that characterize the smart society. However, structures, dynamics and symbolizations remain in the smart politics 'implicit' (*implicit smart politics*), that means not clearly recognizable and not directed towards a process of public and political institutionalization. As we shall see, this situation is linked to the great transitional phase of modern politics (Habermas 1998; Hobsbawn 2007; Reinhard 2007).
2   The lack of institutionalization of the smart politics in genuinely democratic forums entails the risk of forming oligarchies or monopolies and of producing an institutionalization of power in favor of the economic-financial and technological dimensions.

With reference to these theses, the first paragraph analyzes the *implicit smart politics* and its link with some of the main critical points of contemporary politics and of modern state in Western countries, that is to say the relation between civil society and political society and that between representative democracy and direct democracy; the role of political personnel; the role of the national institutions in facing the challenges of the sub-national and supra-national dimension. The second paragraph focuses on the main risks for the democratic development of the smart society and provides some reflections on possible future scenarios. The third paragraph briefly refers to the topic of 'smart-development' in public institutions and affirms the need for a long-term political vision.

Politics and smartness: does 'smart politics' exist?

The concept of 'smartness' has been observed from different perspectives. Literature has often linked the word 'smart' to technological, economic, organizational and communicative development, focusing mainly on the urban sphere (smart city) and linking the idea of smartness to aspects of mobility, general livability, functionality and efficiency (Deaking and Al Waer 2014; Etezadzadeh 2016; Papa and Fistola 2016; Vinod Kumar 2017). Smartness has also received attention in the world of sociology, especially in relation to the discussions around the city and the birth of a post-Fordist and post-modern society (Harvey 2010; Soja 2007), to those on the bond between post-industrial society and smart people (Gurashi 2019) as well as those on relational, identity, and belonging dynamics in the 'smart-community' (Iannuzzi 2019). In the political field, a large part of the literature has often paid close attention to the study and development of smart governance models, without however considering closely the dynamics of power that are a physiological part of them and the process of legitimation on which they are founded (Iannone 2007a: 70). This line of research focuses, therefore, on administrative and managerial aspects (Meijer and Rodríguez Bolívar 2016: 404) and its purpose is to support the creation of smart administrative structures able to adapt to technological development and to contribute to implement smart policies that guarantee the liveability of cities and an improvement in the quality of life. In other words: "Its main focus falls on the capacity of enabling smart solutions through technological patterns at an urban level" (Marciano 2013: 8).

In comparison to the great success of this approach, the studies on the political dimension of 'smartness' are few. Politics seems often to have a secondary role when the word 'smart' comes into play. Albert Meijer and Manuel Pedro Rodríguez Bolívar (2016: 404) write:

> Most publications frame smart city governance as a technical or managerial issue. The underlying assumption is that a smart city makes life better for everyone and there is a lack of attention to the politics of technical choices.

Such a trend is clear, for example, if one compares the studies on the (not smart) city and those on the smart city. While the former are particularly attentive to the dynamics of power present in the urban dimension (Parker 2011; Simmie 2001), to the forms of democracy and citizenship existing in and through the city (Isin 2000) or to the possible politicization of cities (Lanz 2016), among the latter there is a focus on technical practice (Vinod Kumar 2015), economic development (Vinod Kumar 2017) or strategic urban planning and governance (Meier and Portmann 2016). The reason for the lack of interest in politics can be identified in the widespread idea that smartness is a post-political concept (Wiig and Wyly 2016: 487). There are many aspects that have long supported this idea. Firstly, the technical and technological development linked to the word 'smart' has remained for years strongly linked to disciplines not directly characterized by a political dimension (computer studies, engineering, etc.). This has supported the idea of an 'apolitical smartness' in the name of an alleged 'technical neutrality'. Secondly, some of the smart issues themselves have contributed to the myth of post-politics: it is the case of environmental protection, sustainable development and improvement in the quality of life (goals often presented as 'good common' and politically neutral) and/or of themes related to the micro-social dimension and therefore usually less influential from a political point of view – think for example of the smart home (Sessa 2019). This post-political approach has the merit of indicating in the smartness and the idea of smart society elements of novelty with respect to the idea of politics related to the state, to the traditional left-right division or to the undisputed centrality of the political parties in political life. Sure enough, smartness is often far from the central State, difficult to place on the right or on the left and often not directly linked to political parties. However, politics is something broader: it is a dynamic intertwining of interactions, de facto structures and force relations on the one hand and attributions of meaning capable of develop processes of legitimization or delegitimization on the other (Burdeau 1979; de Ghantuz Cubbe 2015; Mongardini 2011), Now, unless we want to affirm that a smart society, just because it has the adjective 'smart', has nothing to do with power structures, factual interactions, power relations, meaning attributions and forms of legitimization and delegitimization, it is rather difficult to deny a connection between politics and smartness. On the contrary, where a smart society exists, there is also a smart politics.

However, smart politics represents a novelty that is not yet stable, whose material and symbolic forms remain unclear. It is therefore an *implicit smart politics*, gradually establishing itself, not yet clearly recognizable and not yet directed towards a clear process of political institutionalization. This situation of transition exists since smart politics is a product and an 'accelerator' of the great transitional phase that characterizes modern politics, largely linked to the processes of globalization and to the transformations of the role and meaning of the modern state (Habermas 1998; Hobsbawn 2007; Reinhard 2007). Hereinafter are four areas in which the link between smart politics and this transitional phase expresses itself.

The first area is that of the relation between civil society and political society. Civil society has been and still is understood in different ways. It is the realm of natural affections and sociability of Adam Smith, but also of the market relations criticized by Karl Marx; it is the realm of the community, but also of individuality, it is a source of legitimacy and stability of political institutions, but also a center of resistance and opposition against them (Rosenblum and Post 2002: 1). However, the different interpretations of civil society seem to have something in common. They usually intend civil society as separated from politics. Civil society, that is, is the society of the relations and dynamics that exist outside the political institutions and outside the 'political society' (Bobbio 2016: 897; Mongardini 2011). In this perspective, politics, gathered in precise institutions and in particular in the state and in Parliament, has the task of representing and synthetizing social plurality. Starting from the 20th century, however, this differentiation between civil society and political society began to fade with mass society and mass parties and with the growing politicization of social life (a process that found its extreme drift in the totalitarian identification of state and society in the 21th century). Nowadays, the extreme pluralism of contemporary societies continues in many respects to weaken the distance between civil society and political society. The different interests of civil society do not let themselves be easily gathered and represented within party or state institutions and demand their own role in political decisions. Political power, therefore, *spreads* among the various social actors (finance, banks, consumers, associations, stakeholders). Thus, a sort of 'organic governance' (a form of de facto direct democracy (Burns 1994)) takes place as well as a liquefaction of decision-making and executive structures (Eckardt 2016: 161), which no longer correspond exclusively to state and party structures. The (idea of a) smart society, together with its political implications, is in line with these processes. In the smart society, citizens, stakeholders and institutions collaborate in the name of shared interests, share decision-making power and defend the interests at stake. Civil society and political society are therefore merged, as the representation of interests as well as the solvability of social conflicts is sought not in an institutionalized relation with a higher authority – as Parliament could be – but in the direct confrontation between the different parts involved. The (idea of a) smart society, therefore, accelerates in this sense the attenuation of the difference between civil society and political society.

The second area is that of the relation between representative democracy and the various forms of direct democracy. The aforementioned attenuation of the distance between civil society and political society and the liquefaction of decision-making and executive centers has gradually brought politics outside the parliamentary arena. The vertical/hierarchical power present in the representative democracy and expressed in the election of the political representatives is sided nowadays by a horizontal and participatory power (that is, not parliamentary). This situation is well summarized by conceptualizations such as those of deliberative and/or participatory democracy, whose central goal is the search for an adaptation between the traditional forms of representation and the needs for greater democratic participation. Hereafter is what Umberto Allegretti (2010: 6–7, my translation) writes of participatory democracy:

A first identification of the notion [...] leads to see participatory democracy, by difference from the more consolidated notions with which it borders – representative democracy and direct democracy –

as a sort of intermediate entity between them and which intersects with them. Sure enough, it represents an interaction, within public procedures – above all administrative, but also legal – between society and institutions, which aims at reaching, through both collaboration and conflicts, at producing a unitary result from time to time, attributable to both these subjects.

The similarity of this form of democracy with the political discourse implicit in the idea of the smart society is clear. Alongside the vote for (parliamentary) representation, smart politics places direct participation in political decisions (e.g. cooperative governance), giving rise to the interaction and collaboration between social actors and public institutions mentioned above by Allegretti and thus flanking vertical/hierarchal power with a horizontal and participatory one. In this perspective, the consequences of the smartness for the traditional institutions are clear also in the idea of 'distributed intellicence' ('distributed' *beyond* public institutions), according to which new technologies, such as social media, open data systems or the network enable social actors to effectively collaborate in political planning and produce solutions for improving quality of life (Meijer 2016: 77). In this context there is also the possibility of *marginalization* of public institutions, as the main role can often be left to "social entrepreneurs [who] are engaged in public problem-solving without any government involvement" (Meijer 2016: 77).

The third area concerns the role of political personnel. Political personnel have often been defined in relation to political parties. According to the phases of evolution of the latter – structure, organization and relation with civil society and the electorate – political personnel have changed their face and function. A classic example is the distinction between "Honoratiorenparteien" and "Massenparteien" (Weber 1919), the former characterized by a non-professional political personnel, recruited among the notables of society, the latter endowed with a complex bureaucratic apparatus, aimed at seeking the consent of those entitled to vote and made up of a class of professionals. Moreover, for Roberto Michels (1910), the processes of bureaucratization and professionalization played a central role in the history of modern political parties (and therefore of political personnel) and the development of the party-organization led inevitably to the formation of a party oligarchy. Schumpeter (2018) stated that the members of a political party were subject in agreement to carry out the competitive struggle for political power. Others considered parties in a more neutral manner as political groups presenting themselves in elections with their own candidates for public office (Sartori 1976), while supporters of the rational choice theory emphasized the action and strategy of individual members, conceiving parties as 'teams' of politicians in the running for the conquest of public government offices (Downs 1957). In these and many other interpretations, political personnel are therefore represented as a *specific* group, with certain political aims, with a precise position within a party organization and then destined to become the holder of a public office. However, starting from the second part of 21st century both the political party and the political personnel have been in a wide-ranging transition, which seems to reduce their 'specificity' as organizations and groups in their own right. New actors have entered politics, or, vice versa, politics has spread to new actors. The birth of new political organizations in the second half of the 21st century has given rise to political forms that were straddling the conceptual space between party and movement (Massari 2007: 111; Gunther and Diamond 2001) and the unbridled spread of political movements, non-governmental organizations and various lobbies of power have brought politics (also) outside of that party oligarchy underlined by much of classical political science. Politics has therefore moved away from being a profession, while political personnel are losing their identity as a class of their own (as we shall see, the processes of professionalization 'moved' from the political to the managerial and technical dimension (see below)). In this context, the 'politician' is now the lobbyist, the entrepreneur or even the simple citizen. This process of enlargement of the political actors includes civic initiatives, forms of urban self-organization, committees, movements, political initiatives of stakeholders. The smart society is no exception. In it, everyone becomes potentially a politician. Citizens are politicians of their own neighborhood and city, stakeholders are politicians of their own interests and institutions are reduced to instruments suitable to facilitate the agreement between the various social parties. In conclusion, the resolution of conflicts (central element of democratic politics) is not necessarily sought among the subjects elected in the representational system, but among the different directly interested parties, who constitute what one might call the new 'pseudo-political personnel', whose characterization remains vague and unclear.

The fourth area is that of the relation between the central and local institutions. The tension between multiple levels of governance (supra-national, national, sub-national) has by now become a classic theme of political science. On the one hand, the processes of globalization become even stronger, placing the role of the state in a phase of transition, or of crisis (Habermas 1998; Hobsbawn 2007; Reinhard 2007). On the other hand, forms of nationalism and sovereignism ferment, as well as forms of regionalization or devolution processes (as in the Scottish case), instances of independence (as in the Catalan case), or, more generally, institutional and regulatory tensions between the central and local administration, which led for example to long discussions in Germany (Zei 2012). There is, therefore, a vertical tension between a 'center', an 'above' and a 'below', that means between the state and the authorities superordinate to it and subordinate to it. The smart society is fully part of these processes, and in particular of those between the 'center' and the 'below'. Smartness means participation, interconnection, sharing of data, interests and decision-making power between actors involved in the same level (mainly the urban or suburban level, which favors participation and interconnection thanks to the presence and involvement on the territory) and united by the same interest for a common goal. Smart politics, therefore, takes place mainly horizontally and contrasts – at least potentially – with the hierarchical power that characterizes the political-institutional relation between different administrative levels, namely between central and local administration, which represent different interest, goals and priorities.

In this context, some open questions remain. First, the form and the dynamics that characterize the smart politics remain unidentifiable. Political authority, decision-making power and the abovementioned 'organic governance' remain implicit and contingent and often exist just de facto and not de jure. From a political-institutional point of view, it is not yet clear how the various intermediate forms of democracy (neither representative nor direct) that develop gradually in the smart society will integrate with the traditional representative system, nor is it clear what specific (new) institutional forms will be created. In the same way, political personnel remain indefinable, while political power remains widespread among the various social actors.

Politics, smartness and undemocratic trends

In the previous paragraph we tried to show how much political potential characterizes the smart society and how many political implications it involves. More precisely, we wanted to affirm the existence of an *implicit smart politics*, understood as a political vision, as a set of dynamics, ideas and representations linked to the concept of smartness and to its relation to the great changes of modern politics. We then identified as the main criticalities of the smart politics the strong diffusion of power among social actors, the difficulty of recognizing and controlling power and the absence of a real process of political institutionalization of the latter. In this context, a central risk factor for a genuinely democratic development of smart politics is the concentration of political power in the hands of the economic-financial and technological-communicative organizations.

They may become – and in part already are – the *implicit* holders of decision-making power. Regarding the economic dimension, the thesis that we want to support here is that the idea of smartness, together with that of governance, risks promoting an excessive neo-liberalization of society and politics without providing due forms of protection from excessive power of economic-financial actors. Part of the literature has been discussing this for some time. Geddes (2005) writes about the "Overlapping Regimes of Local Governance in a Neoliberalising World". Hollands (2008: 314), refers to the risk of a "high-tech urban entrepreneurialism", while other authors intend the "smart city utopia" as a "fundamental facet of the neoliberal contemporary ideology" (Grossi and Pianezzi 2017: 80). Similarly, local governance, so important for the smart society, has been linked to the growing "neoliberalization of urban political-economic space" (Brenner and Theodore 2002: 376) and to the development of an entrepreneurial urban leadership (Painter and Goodwin 2000), mainly made up of managerial staff. Such a process creates thus a new professionalized class (in spite of the abovementioned enlargement and de-professionalization of the political personnel, therefore, takes place in the background a re-professionalization in economic and technical sense) and the management of local politics is often and largely entrusted to private corporations. This is the case of the smart city, defined by Hollands (2015) as a "corporate smart city". In it, the introduction of market elements in the local policies and the logic of demand/supply aimed at the needs of consumption show no sign of diminishing.

Defining such a situation as necessarily negative would perhaps be exaggerated. The mechanisms and the logic of the market are not necessarily dangerous. They can work as a support for the public administration thanks to their greater financial resources and to 'expertise' that are often not available in the context of public institutions. Risks for the democratic society and for the welfare state still exist, however, if the logic of the market does not encounter restraints. The first risk is that "this business-driven development of smart city might result in the prioritization of business goals over social and economic ones, thus leading to social polarization and inequality" (Grossi and Pianezzi, 2017: 79). Without adequate counter-balances, indeed, the market tends to reproduce and increase already existing disproportions and advantages (Rössler 2015). Parker (2011: 114), similarly, stresses the possible risk of "the privileging of market values and activities over other more traditional local government functions and priorities (such as the provision of local welfare services and the provision of locally consumed amenities)". Without restraints, therefore, economy turns into *economism* (Mongardini 2007), which is an ideology that reduces society and politics to economic terms. In other words, economics may become the center of political life and make smart politics slip away from the principle of legitimacy, the supporting base of every democracy. At the same time, the institutions, deprived of their own authority and their role, can only "serve the cause of progress, even if this were to be controversial" (Iannone 2007b: 61–62).

Similar dangers also come from the technological dimension. Beyond the myth of the neutrality of technology, one should affirm that digital devices "are not just technologies, understood as neutral instruments through which information is circulated. They are, above all, devices of meaning, which articulate and set the conditions for new social [and political] practices" (Marciano 2018: 141). In this sense, as several studies have already observed (Bijker et al. 1993; Winner 1980), technological tools are related to the cultural and political context of meaning of a given society. From the point of view of the concrete interactions of power, moreover, they represent the instruments par excellence of the technical power, that is, the power to create facts or usable artifacts (Popitz 2001: 22–23). This is a "materialized power", whose application and effect are mediated by technical objects (Popitz 2001: 136). In the smart society such a power meets a very fertile field for its spread. For example, Vanolo (2013: 39) writes about the smart city:

> Among its various effects, the discourse on smart city tends to give rise to and support a redefinition of urban space aimed at producing and justifying technological hyper-rationalities that design new geometries of power within the city.

Furthermore, the risks deriving from the technical and technological power increase if technology meets the market logic. Vanolo (2013: 39) writes:

> The smart city re-encodes the discourse on city government in a technological-environmental sense, widening the playing field of consultants and private companies. In this sense [...] the vision of the smart city [...favors] more docile and disciplined cities, ready to mate with political-technological assemblages and devices predisposed to naturalize and justify new arrangements for the circulation of capital and its rationales inside the city.

Such a mixture – the mixture between technology and capital – generates a trade of information and interests of incalculable value. This is the case of personal data. They, which are not commercial goods, become a market instrument through a process of commodification (Ertman and Williams 2005), where "commodification means turning something into a commodity in order to exchange it on a market, where that 'something' is usually regarded as an object that should not be so commodified" (Rössler, 2015: 147). These data are used to carry out commercial targeting practices, aimed at directing consumption in an 'intelligent' way, thus identifying age, tastes and habits of users and implementing forms of social sorting, if not discrimination, useful for maximizing profits through targeted advertising (Turow, 2011).

The question arises whether these risks are containable, or whether they constitute a concrete obstacle to the development of a democratic smart politics. Without wishing to reduce the analysis to rigid schematizations, one can try to identify three approximate future scenarios. In the first, one could incur in a strengthening of the already existing economic-financial and technological hierarchies and in the creation of new ones. Without an institutionalization of power in genuinely democratic forums, the implicit holders of decision-making power will be increasingly invasive, influential, able to dictate rules and to affirm their own interests in order to decide on sensitive data and their use. In this sense, power will be more and more apolitical in its form (namely, not expressed in political arenas) but political *in its substance*. In the second scenario, the implicit smart politics would become completely explicit and would develop in a democratic institutionalization process. (Smart) democratic life would take place in political arenas, suitable for recognizing but also containing the spread of political power. In this sense, power would be political both in its form and in its substance. The third scenario, which is the most likely, occupies a middle position between the first and the second. (Smart) power remains 'hidden' and difficult to control, and the democratic life of the smart society does not achieve its clear and stable configuration. However, the various social and political actors gradually develop adjustments and counter-balancing mechanisms to the uncontrollability of the power, producing new regulations and gradually introducing new institutional forms. In such a direction, for example, there is the issue of privacy, in which public institutions, consumer protection organizations and other social actors exert their pressure for the development of advanced legislation aimed at regulating the trade in personal data.

Some short mentions about the institutional praxis: the need for a long-term vision

In recent decades, public institutions have tried to adapt to the labels of smart government or smart governance, mainly through the integration of technological tools in their administrative circuits, aimed at improving communication and cooperation with other institutions, with citizens and with stakeholders, ensuring transparency, increasing overall efficiency and reducing expenses. Exemplary is the so-called e-participation, characterized by the use of ICT technologies to foster and facilitate political participation, or the concepts of 'open governance' or 'open government', according to which public institutions are required to guarantee a high degree of 'openness' and transparency towards citizens. The idea of an open smart city, for example, was formulated in order to indicate all the tools, projects and initiatives aimed at recognizing, managing and satisfying the high need for information and communication necessary for the functioning of urban governance (Habenstein et al. 2016: 56). The variety that characterizes these concepts has given rise to empirical studies (Manville et al. 2014; McLaren and Agyeman 2015), while many local administrations have undertaken concrete attempts at smart experiences (Amsterdam, Barcelona, Copenhagen, Vienna). What nevertheless emerges is that the current approach is often based on a functionalist and efficient perspective, which fails to support the emergence of an overall political vision. Concepts such as 'smart governance' usually lose their theoretical meaning and their symbolic-ideological dimension (Iannone 2005: 63) and are too often operationalized and considered uncritically. The realization of a democratic smart politics requires, on the contrary, a political reflection that goes (also) beyond the search for efficient administration.

The political dimension of the smart society must clearly come to light and take the form of democratic politics (that is, democratic conflict and balance between powers and counterpowers). The system of exchange and cooperation between citizens, stakeholders and institutions must begin to seek more clearly a departure from its contingent and situational form. To this end, it could be useful to further support the creation of regular meetings dedicated to *smart development*, the strengthening of long-term projects (this is the direction, for example, of the European Union with Horizon 2020 – Work Program 2018–2020) (Gurashi, Iannuzzi and Sessa 2019), as well as a more precise definition and organization of data sharing and participation in decision-making processes. Furthermore, it is necessary to create clearer control tools, which prevent the formation of oligarchies and monopolies. A more in-depth discussion of the processes and sources of legitimization of the various forms of governance should also be fostered (Iannone 2016: 65). Moreover, in regard to the relation between the central and local institutions, the continuous elaboration of different smart projects without a comprehensive political plan and the often-unclear distribution of competences between central and local institutions, risk causing new bureaucratic complications. Also in this case, therefore, new discussions on long-term projects and structural innovations are needed. A redefinition of the role of the principle of subsidiarity could be a starting point. In this sense, the institutionalization of general conferences between the local entities of a given territory, aimed at coordinating technological development and smart design, should be supported.

Conclusion

As we have tried to suggest, the smart society is characterized by a strong political potential. The *implicit smart politics* is linked to the great questions and major changes in modern politics, such as the relation between political society and civil society, that between representative democracy and direct democracy, or the role of political personnel and that of the modern state in facing the challenges of the sub-national and supranational dimension. 'Smartness' therefore also means 'politics', that is, a new politics that has not yet found its own structure. On the one hand, this politics presents itself in a democratic form. On the other hand, it carries with itself strong risks for the development of a genuinely democratic political life. The main danger is that of an excessive concentration of power in the hands of the economic-financial and technological world. Faced with this risk, it is necessary that the discourse around the concept of smartness definitively leaves the myth of its alleged unpolitical nature. Political Science can provide great support at this stage. To this end, it must look beyond the classic topics of the parliamentary arena, political parties and electoral systems. Political discourse, like politics in general,

> must adapt to the new spaces of collective life and necessarily pass through an anarcho-constitutive phase of a new type of organization, with respect to which the function of national states will necessarily be different and partly conditioned by the new realities. New realities, [...] that find their primary engine in the development of capitalism, which transforms culture and politics.
>
> (Mongardini 2007: 110)

References

Allegretti, U. (2010) "Democrazia partecipativa: un contributo alla democratizzazione della democrazia." In Allegretti, U. (ed.) *Democrazia partecipativa: esperienze e prospettive in Italia e in Europa*. Firenze: Firenze University Press, 5–45.

Bijker, W. E., Hughes, T. P. and Pinch, T. (eds.) (1993) *The Social Construction of Technological Systems: New Directions in the Sociology and History of Technology*. Cambridge, Massachusetts: MIT Press.

Bobbio, N. (2016) Società civile. In Bobbio, N., Matteucci, N. and Pasquino, G. (eds.) *Dizionario di politica.* Novara: De Agostini Libri, 893–897.

Brenner, N. and Theodore, N. (2002) "Cities and the Geography of 'Actually Existing Neoliberalism'." *Antipode* 34(3), 349–379. doi:10.1111/1467-8330.00246.

Burdeau, G. (1979) *Le libèralisme.* Paris: Editions du Seuil.

Burns, T. (1994) "Post-Parlamentary Democracy. Sacralities, Contradictions and Transitions of Modernity." In Mongardini, C. and Ruini, M. (eds.) *Religio. Ruolo del sacro, coesione sociale e nuove forme di solidarietà nella società contemporanea.* Roma: Bulzoni.

Cladis, M. (2010) "Religion, Secularism and Democratic Culture." *The Good Society* 19(2), 22–29. doi:10.1353/gso.2010.0012.

de Ghantuz Cubbe, G. (2015) "La base ideologico-progettuale dello stato sociale." *Rivista trimestrale di scienza della amministrazione* 59(2), 24–38. doi:10.3280/SA2015–002002.

de Ghantuz Cubbe, G. (2017) "Radical Secularism and the Risks of Undemocratic Trends. Constitutional States, 'Far Right' Parties and Anti-Immigration Platforms." *Rivista trimestrale di scienza dell'amministrazione* 61(2), 1–21.

Deaking, M. and Al Waer, H. (eds.) (2014) *From Intelligent to Smart Cities.* London: Routledge.

Downs, A. (1957) *An Economic Theory of Democracy.* New York: Harper & Row.

Eckardt, F. (2016) "Urban Governance und „e-Participation"? Innovative Politik in der medialisierten Stadt." In Behrens, M., Bukow, W. D., Cudak, K. and Strünck, C. (eds.) *Inclusive City: Überlegungen zum gegenwärtigen Verhältnis von Mobilität und Diversität in der Stadtgesellschaft.* Wiesbaden: Springer VS, 161–172.

Ertman, M. M. and Williams, J. C. (eds.) (2005) *Rethinking Commodification. Cases and Readings in Law and Culture.* New York: New York University Press.

Etezadzadeh, C. (2016) *Smart City – Future City? Smart City 2.0 as a Livable City and Future Market.* Wiesbaden: Springer Vieweg.

Geddes, M. (2005) "Neoliberalism and Local Governance – Cross-National Perspectives and Speculations." *Policy Studies* 26(3–4), 359–377. doi:10.1080/01442870500198429.

Grossi, G. and Pianezzi, D. (2017) "Smart Cities: Utopia or Neoliberal Ideology?" *Cities* 69, 79–85. doi:10.1016/j.cities.2017.07.012.

Gunther, R. and Diamond, L. (2001) Types and Functions of Parties. In Gunther, R. and Diamond, L. (eds.), *Political Parties and Democracy.* Baltimore: Johns Hopkins University Press, 3–39.

Gurashi R. (2019) "Smart people: The Individual Challenge of the Fourth Industrial Revolution." In Iannone, R. and Gurashi, R. with Iannuzzi, I., de Ghantuz Cubbe, G. and Sessa, M., *Smart Society. A Sociological Perspective on Smart Living.* Abingdon: Routledge.

Gurashi, R., Iannuzzi, I. and Sessa, M. (2019) "From theory to empiricism: the challenge to the future of the Sapienza University at Solar Decathlon Middle East." In Iannone, R. and Gurashi, R. with Iannuzzi, I., de Ghantuz

Cubbe, G. and Sessa, M., *Smart Society. A Sociological Perspective on Smart Living*. Abingdon: Routledge.

Habenstein, A., D'Onofrio, S., Portmann, E., Stürmer, M. and Myrach, T. (2016) "Open Smart City: Good Governance für smarte Städte." In Meier, A. and Portmann, E. (eds.) *Smart Cities. Strategie, Governance und Projekte*. Wiesbaden: Springer Vieweg, 47–71.

Habermas, J. (1998) *Die postnationale Konstellation. Politische Essays*. Frankfurt am Main: Suhrkamp.

Harvey, D. (2010) *La crisi della modernità*. Milano: Il Saggiatore.

Hobsbawn, E. J. (2007) *La fine dello stato*. Milano: Rizzoli.

Hollands, R. G. (2008) "Will the Real Smart City Please Stand Up?" *City* 12 (3), 303–320. doi:10.1080/13604810802479126.

Hollands, R. G. (2015) "Critical Interventions into the Corporate Smart City." *Cambridge Journal of Regions, Economy and Society* 8(1), 61–77.

Iannone, R. (2005) "Governance: una questione di significati." *Rivista trimestrale di Scienza dell'amministrazione* 2, 59–87.

Iannone, R. (2007a) "La società delle reti. Cerchie, capitale sociale e governance." *Rivista trimestrale di scienza dell'amministrazione* 52(3), 28–106.

Iannone, R. (2007b) *Società dis-connesse. La sfida del digital divide*. Roma: Armando.

Iannone, R. (2016) "Network Society. What is it?" In Iannone, R., Ferreri, E., Marchetti, M. C., Mariottini, L. and Cipri, M., *Network Society. How relations rebuild space(s)*. Wilmington, Delaware: Vernon Press, 21–75.

IannoneR. (2019) "Smart Society. The critical sense of a world strategy." In Iannone, R. and Gurashi, R. with Iannuzzi, I., de Ghantuz Cubbe, G. and Sessa, M., *Smart Society. A Sociological Perspective on Smart Living*. Abingdon: Routledge.

Iannuzzi, I. (2019) "Smart community: A New Way of Being Together?" In Iannone, R. and Gurashi, R. with Iannuzzi, I., de Ghantuz Cubbe, G. and Sessa, M., *Smart Society. A Sociological Perspective on Smart Living*. Abingdon: Routledge.

Isin, E. F. (ed.) (2000) *Democracy, Citizenship and the Global City*. London: Routledge.

Lanz, S. (2016) "Politik zwischen Polizei und Post-Politik: Überlegungen zu , urbanen Pionieren' einer politisierten Stadt am Beispiel von Berlin." In Behrens, M., Bukow, W. D., Cudak, K. and Strünck, C. (eds.) *Inclusive City: Überlegungen zum gegenwärtigen Verhältnis von Mobilität und Diversität in der Stadtgesellschaft*. Wiesbaden: Springer VS, 43–61.

Manville, C., Cochrane, G., Cave, J., Milliard, J., Pederson, J. K., Thaarup, R., Liebe, A., Wissner, M., Massink, R. and Kotterink, B. (2014) "Mapping Smart Cities in the EU." http://www.europarl.europa.eu/RegData/etudes/etudes/join/2014/507480/IPOL-ITRE_ET(2014)507480_EN.pdf (March 27, 2019).

Marciano, C. (2013) "Unpacking a Smart City Model." *International Journal of Interdisciplinary Environmental Studies* 7(3), 1–12.

Marciano, C. (2018) "Democrazia locale e media digitali. Verso una critica della Smart City." In Vitale, E. and Cattaneo, F. (eds.) *Web e società democratica: un matrimonio difficile.* Torino: Accademia University Press, 141–156.

Massari, O. (2007) *I partiti politici nelle democrazie contemporanee.* Roma: GLF editori Laterza.

McLaren, D. and Agyeman, J. (2015) *Sharing Cities: A Case for Truly Smart and Sustainable Cities.* Cambridge, Massachusetts: MIT Press.

Meier, A. and Portmann, E. (eds.) (2016) Smart Cities. Strategie, Governance und Projekte. Wiesbaden: Springer Vieweg.

Meijer, A. (2016) "Smart City Governance: A Local Emergent Perspective." In Gil-Garcia, J. R., Pardo, T. A. and Nam, T. (eds.) *Smarter as the New Urban Agenda. A Comprehensive View of the 21st Century City.* Cham: Springer, 73–85.

Meijer, A. and Rodríguez Bolívar, M. P. (2016) "Governing the Smart City: A Review of the Literature on Smart Urban Governance." *International Review of Administrative Sciences* 82(2), 392–408. doi:10.1177/0020852314564308.

Michels, R. (1910) *Zur Soziologie des Parteiwesens in der modernen Demokratie: Untersuchungen über die oligarchischen Tendenzen des Gruppenlebens.* Leipzig: Klinkhardt.

Miccoli, G. (2010) "Sulla storia del concetto di laicità." *Studi e Materiali di Storia delle Religioni* 76(1), 257–266.

Mongardini, C. (2007) *Capitalismo e politica nell'era della globalizzazione.* Milano: Franco Angeli.

Mongardini, C. (2011) *Pensare la politica: per una analisi critica della politica contemporanea.* Roma: Bulzoni.

Painter, J. and Goodwin, M. (2000) "Local Governance after Fordism: A Regulationist Perspective." In Stoker, G. (ed.) *The New Politics of British Local Governance.* Basingstoke: Macmillan, 33–53.

Papa, R. and Fistola, R. (eds.) (2016) *Smart Energy in the Smart City: Urban Planning for a Sustainable Future.* Wiesbaden: Springer.

Parker, S. (2011) *Cities, Politics and Power.* London: Routledge.

Popitz, H. (2001) *Fenomenologia del potere.* Bologna: Mulino.

Reinhard, W. (2007) *Geschichte des modernen Staates. Von den Anfängen bis zur Gegenwart.* München: C. H. Beck.

Rosenblum, N. L. and Post, R. C. (2002) *Civil Society and Government.* Princeton, New Jersey: Princeton University Press.

Rössler, B. (2015) "Should Personal Data Be a Tradable Good? On the Moral Limits of Market in Privacy." In Rössler, B. and Mokrosinska, D. (eds.) *Social Dimensions of Privacy. Interdisciplinary Perspectives.* Cambridge: Cambridge University Press, 141–161.

Sartori, G. (1976) *Parties and Party Systems. A Framework for Analysis.* Cambridge: Cambridge University Press.

Schumpeter, J. A. (2018) *Kapitalismus, Sozialismus und Demokratie.* Stuttgart: UTB.

Sessa, M. (2019) "Home smart home. A sociology of living in the age of reflective materialism." In Iannone, R. and Gurashi, R. with Iannuzzi, I., de

Ghantuz Cubbe, G. and Sessa, M., *Smart Society. A Sociological Perspective on Smart Living*. Abingdon: Routledge.

Simmie, J. (2001) "Planning, Power and Conflict." In Paddison, R. (ed.) *Handbook of Urban Studies*. London: Sage, 385–401.

Soja, E. W. (2007) *Dopo la metropoli. Per una critica della geografia urbana e regionale*. Bologna: Patròn.

Turow, J. (2011) *The Daily You. How the New Advertising Industry Is Defining Your Identity and Your Worth*. New Haven: Yale University Press.

Vanolo, A. (2013) "Smart city, condotta e governo della città." In Santangelo, M., Aru, S. and Pollio, A. (eds.) *Smart city. Ibridazioni, innovazioni e inerzie nelle città contemporanee*. Roma: Carocci, 39–52.

Vinod Kumar, T. M. (ed.) (2015) *E-Governance for Smart Cities*. Singapore: Springer.

Vinod Kumar, T. M. (ed.) (2017) *Smart Economy in Smart Cities. International Collaborative Research: Ottawa, St. Louis, Stuttgart, Bologna, Cape Town, Nairobi, Dakar, Lagos, New Delhi, Varanasi, Vijayawada, Kozhikode, Hong Kong*. Singapore: Springer.

Weber, M. (1919) *Politik als Beruf*. München: Duncker u. Humblot.

Wiig, A. and Wyly, E. (2016) "Introduction: Thinking Through the Politics of the Smart City." *Urban Geography* 37(4), 485–493. doi:10.1080/02723638.2016.1178479.

Winner, L. (1980) "Do Artifacts Have Politics?" *Daedalus* 109(1), 121–136.

Zei, A. (2012) "Chi governa in Germania? Le riforme contese tra Berlino, Karlsruhe, Francoforte e Bruxelles." In Lanchester, F. (ed.) *La Costituzione degli altri: dieci anni di trasformazioni in alcuni ordinamenti costituzionali stranieri*. Milano: Giuffrè, 95–134.

3 Smart people and prosumers

The individual challenge to the Fourth Industrial Revolution

Romina Gurashi

Introduction

In contemporary society, intelligence has become an essential qualifying element not only for the individual as such but also for the community to which he belongs, for the way in which he administers his personal or collective assets (de Ghantuz Cubbe 2019), because of the way he lives and of the technology he uses. Being *smart* has become a cross-cutting need for multiple fields of human life. But what does being a *smart person* mean? Does it mean being of above average intelligence or simply having skills that allow you to use technologies efficiently? If one is a *smart person* can he also be a *prosumer*? And finally, when do individuals cease being consumers to become *prosumers*?

These are just some of the questions that this chapter wants to address, aimed at trying to offer an overview of the increasingly pervasive role of technology in all areas of human life in this precise phase of knowledge capitalism. Technologies are changing the way individuals experience not only their individuality but also their relational dimension, which is now firmly embedded in global networks that crumble the traditional community dimension, building a new one (Iannuzzi 2019) based on use of mediation technologies, weak ties, contingency, deterritorialization, individualism and hedonism.

If, therefore, technology and science have offered tools able to improve the lives of smart people and enhance their working and interactive abilities, they are also producing important criticalities linked to the limits of science and technology, to the perverse effects of hyper-connection, of hedonistic consumerism typical of post-industrial society as well as the liquefaction of social relations and the rise of uncertainty linked to the speed of propagation of risks and global catastrophes.

Technology is rapidly turning the face of the most advanced capitalistic societies into 'smart societies' (Iannone 2019) where the integration between man and machine is leading to the emergence of new forms of social interaction (which are developed through graphic and digital interfaces and through internet) and new social actors (*smart people*, robots and humanoids) that manifest new needs and ask for new tools to satisfy them.

The industrial world is therefore trying to identify new ways and new tools to satisfy these needs by producing changes that come from the world of the factory and invade society in all its many dimensions. After pointing out that technology offers undeniable benefits in terms of improving lifestyles and opportunities for access to knowledge, experience and assets, we need to ask ourselves what the perverse effects of this change are and whether the balance tends more towards a negative or positive result in terms of sustainability.

The post-industrial society and the *smart people*

In the last few years, in the academic world and in the world of professions, there has been increasing talk of the Fourth Industrial Revolution and Industry 4.0. Both names have been used to indicate the same historical, economic, political and social phenomenon, i.e. a new phase of advanced capitalism characterized by the introduction of digital innovations, automatisms and networks within industrial production systems. While the definition of 'Industry 4.0' has been taken over by the world of professions and focuses specifically on changes in the industrial production system, the definition of Fourth Industrial Revolution has been used more often in the academic world and precisely in the context of social sciences to indicate not only a change in the ways of production, but also the changes that the introduction of automations, big data, internet of things, cloud computing, robotics, digital networks have involved in living socially, in changing lifestyles, in changes in the taste and consumption choices.

The integration between the physical and digital world favored by the development of Information and Communication Technologies has opened up new industrial scenarios to the industrial world that apply the idea of networks also to machines and factories (Kengermann, Lukas and Wahlster 2011). They are now included in a system of global connections that enables them to use intelligent products that are autonomously capable of recognizing any supply chain problems, and consequently reorganize the problem (Posada, Toro and Barandiaran 2015: 2–3) both from a process reorganization and from an error resolution point of view. The *Internet of Things* (IoT) enters factories through *smart manufacturing,* i.e. through the simultaneous adoption of both digital technologies capable of increasing the interconnection and cooperation of resources (machinery, products, people and information) both in the internal production sequence within the factory, and along the value chain processes.[1] These are technologies that can be divided into two large groups: the first, inherent to the actual Information Technologies (IT) and formed by the Internet of Things, Big Data and Cloud Computing, and the second, more heterogeneous and close to the operational level, formed by Advanced automation, Advanced Human Machine Interface (HMI) and Additive Manufacturing.

Particularly interesting are the Advanced HMIs, that is wearable devices and graphic interfaces that can favor communication between human being and machine and are used to acquire or send visual, sound and/or tactile notifications. These are devices whose predominant function is to acquire data, which are used in factories to monitor or implement the employee's work functions but are also increasingly used outside the working life to obtain information on health status (fitness tracker and smart watch), as an extension of one's body (acoustic devices and action cams) or as tools to project oneself into a virtual reality (digital viewers).

The HMI is also at the center of one of today's most important industries in terms of breadth and future growth prospects: the *Internet of Services* (IoS). The advent of the Internet has in fact allowed the diffusion of offer and consumption of services of all kinds, from eBooks, to online consultation of one's financial situation, to video calls with friends and distant relatives (Soriano, Heitz, Hutter et al. 2013: 284), to the provision of services to the citizen by municipalities through digital platforms, to cloud storage of photos and documents and so on. The economy of the IoS is based on the creation, production and sale of 'intangible products' whose use takes place through the increasingly intensive use of electronic devices. The use of mobile devices by end users, i.e. *smart people*, is leading to an idea of the Internet that is new and radically different from the past. Here "the human user becomes more central than ever, [...] their personal devices become their proxies in the cyber world, in addition to acting as a fundamental tool to sense the physical world" (Conti, Passarella and Das 2017). The main effect of this paradigmatic change of advanced capitalism is that will to put the individual at the center of studies on Science Technology and Society (STS), confirming the idea proposed by Kuhn (1962) for which *scientific facts are products whose origin lies in socially conditioned scientific research and whose technological applications are often found in relation to other forms of social development in the juridical, political, ethical, and cultural sphere, which in the current capitalist context are rooted in the development of global market.*

This discussion of the foundations of the paradigm of classical capitalism[2] implemented starting from the 1970s marked the transition from the industrial economy based on *production* to the post-industrial economy based on *knowledge* (Antonini 2006: 18).

Within the debates on STS, the post-industrial economy and the post-industrial society have been described in two radically different ways, which however have had in common the ambition to highlight aspects of the industrial revolution currently going on.

According to Daniel Bell, for example, post-industrial society is a hyper-industrialized society characterized by the consolidation of the link between science, technology and work organization for the expansion of innovation processes not so much in the secondary sector as in the tertiary sector. Scientific advances and technological developments fuel the processes of hyper-capitalism and pervade every area of human society, up to making *techné* "a form of art that bridges culture and social structure, and in the process reshapes both" (Bell 1991: 20).

By contrast, Alain Touraine insisted on the transformative capacities of society and the discontinuities between industrial society and post-industrial society rather than on the acceleration of trends already evident since the dawn of capitalism. Criticizing structural-functionalism, Touraine claimed the "incapacity to understand the constant transformation of society by social actors, since it sees the latter merely as the manifestations of a hidden domination" (Touraine 2002: 388).

Bell and Touraine therefore agreed in recognizing that a new type of society was being formed, based at the same time on the change of the occupational structure and social roles as well as on the disuse of social categories and institutions such as state, nation, democracy, class and family that helped imagine and build society (Touraine 2013), but while Bell highlighted the importance of industrial technological progress, the second (Touraine) overplayed his hand on the overcoming of a capitalism based on the distribution of wealth in favor of a capitalism based on the cultural definition of individual needs.

The birth of this new type of society and the overcoming of the idea of the transformative power of the class has meant the weakening and partial destruction of the idea of society as a determinant of the idea of good and evil in terms of social utility. This marked the transition to an individuality that goes beyond the limits of social life to make room for the expression of one's own personality through one's corporality and actions (Touraine 2013: 395).

If therefore the individuals and the technologies they employ are profoundly transforming the society in which we live, it is more important than ever to ask ourselves with more and more scruple: *who are the smart people? And, where is this race for technological development taking us?*

The current historical-social phase is characterized by processes of permanent innovation that require higher levels of training, continuous learning skills and specific skills that imply the need for an individual propensity for adaptability, mobility, and flexibility. This means that the social actor – in this case the *smart person* – is characterized by the possession of higher levels of qualification (and therefore of skills, abilities and knowledge and not only of high levels of education), propensity to learning, open-mindedness and creativity. For historical, statistical and generational reasons, the cluster that nowadays best represents this social category is that of the *Millennials*. They, unlike Generation X that preceded them, are in fact a generation that was born in the period of the advent of technology and grew up in parallel with its development[3]. For this very reason this category of people has spontaneously acquired the rules of digital communication (Kamensky 2017) by learning to exploit them more effectively than their parents. The age of the group – which is between 39 and 19 – combined with the technological and IT skills developed during growth, make this generation the most active in promoting scientific, technological and social change. Generation Z, which is made up of individuals born between 1999 and 2010 (Twenge 2017) is, in fact, too young to be fully aware of the aims and possible critical use of the digital tools at its disposal. The Millennial generation is also characterized by being an inquiring generation with respect to the problems of ethical consumption, environmental sustainability and the fight against inequalities. A generation for which '*risk*' represents such an important analytical category to systematically condition the ways they face the problems and the insecurities induced and introduced by modernization. As highlighted by Ulrich Beck "the risks, as opposed to more ancient dangers, are consequences related to the threatening force of modernization and its globalization of doubts" (Beck 2005: 21). *They are the product of scientific and technological advances determined by an industrial advancement tending to overproduction that makes the dangers no longer limited in time and space, but so global as to be difficult to ascribe to any individual or entity.* The risk and uncertainty of contemporary societies are so unpredictable both in time and space that they literally collapse the units of measurement identified by science and regulatory institutions (Beck 2005: 22) for determining its size.

The development of *téchne* that has made science, technology and knowledge democratically accessible to all also has a bitter side to the coin that lies in the increase in the economic, political, social and environmental risks associated with globalization, by which the smallest occurrence in a remote area of the world can have negative repercussions even on the other side of the globe.

These uncertainties are examined by studies on STS that increasingly analyze, on the one hand, the relationship of individual and social actions in the development and control of science, technology (i.e. *social construction of technologies*), and on the other, the risks and the opportunities that science and technology can represent for peace, security, community, democracy, environmental sustainability, that is for all those human values that are of fundamental importance for smart people (Latour 1987).

The world of cognitive capitalism: the role of network and connectivity

The advent of cognitive capitalism has led to questioning the traditional concepts of time, space and distance. Life within an ultra-connected world has led individuals to deconstruct the times of life and work that are no longer determined by factory shifts or business clock, but are made up of moments that are part of a 'timeless time' made possible by new types of work flexibility such as *smart working* (Botteri and Cremonesi 2017) that combine to blend and confuse the times of life and work, the times of leisure and those of industriousness. In the network society, space[4] and time are no longer characterized based on the contiguity or geographical distance of one place from another, but on the basis of virtual relationships that allow us to reach a place through the use of information technology (Castells 2009: 33–34). The development of the Internet and social networks places the world at the click of a button, making digital life the new social reality in which time is compressed and not sequenced through the progressive dissolution of past and future into a present made of immediacy.

David Harvey had defined the phenomenon with the term 'space-time compression'. A compression whose meaning was that of expropriating the nature of its cultural value and its spatial attributions in favor of a progressive increase in the speed of circulation of capital and a consequent re-discussion of the social sense of place. "As a space appears to shrink to a 'global village' of telecommunication and a 'spaceship earth' of economic and ecological interdependencies [...] and as time horizons to the point where the present is all there is [...] so we have to learn how to cope with an overwhelming sense of *compression* of our spatial and temporal worlds" (Harvey 1990: 240). Therefore, being here and now becomes more important than tomorrow. An attitude that poses serious questions about the consequences of the actions dictated by the logic of 'here and now', dictated by the need to satisfy contingent needs, and which clashes with the need for an ecological approach to human actions. The space-time compression reduces the distances to a 'global village' entirely viable and reachable through technological tools, but in so doing it also contributes to liquefying traditional social relations, replacing them with new and weaker ones.

In the network society relations play a fundamental role in shaping the global society and the alternation between the *strong ties* and the *weak ties* identified by Mark Granovetter (1973) plays a decisive role in the construction of the experiential dimension of smart people. According to Granovetter, the *strong ties* are established between people who know each other very well, such as family members and very close friends, while the *weak ties* mainly concern acquaintances and superficial friendships. One could therefore think that *strong ties* are qualitatively better and preferable to *weak ties*, but *weak ties* play a very important role in the circulation of information[5] and in increasing the possibilities of individual mobility. "Individuals with few or no weak ties will be deprived of information from distant parts of the social system and will be confined to the provincial news and views of their close friends. This deprivation will not only insulate them from the latest ideas and fashions but may put them in a disadvantageous position" (Granovetter 1983: 202) both within the labor market system, both in participation in political movements and in the participation in goal-oriented organizations. The *weak ties* therefore play an essential role simultaneously for the circulation of information within the network society and for achieving the individual goals of the smart people. However, the decrease in the quantity and quality of *strong ties* due to the decrease in face-to-face communication is helping to reinforce the sense of isolation and solitude that in some individuals has been positively correlated with the development of diseases such as depression, stress and anxiety (Kraut, Patterson and Lundmark 1998).

The circulation of information, the acceleration and the promiscuity in the times of life and work as well as the dissolution of the territorial dimension have had a profound impact on the psychology and the values of the smart people that are today increasingly projected to the instantaneousness of consumption and experiences in a hedonistic attitude that seems uncaring of tomorrow. Being always connected in social and digital networks today has an essential relevance in order to build those utilitarian relations of knowledge and superficial friendship that are essential to the circulation of information and the pursuit of individual goals. In an era in which knowledge and information represent the central value of the economic and social system, being "disconnected, or superficially connected to the Internet is tantamount to marginalization in the global, networked system" (Castells 2001: 269). The issue of unfair access to technology and the inadequate development of the ability to use it between the North and South of the world, between capitalistically advanced countries and underdeveloped countries is producing an unbridgeable digital divide, so that a part of the world is cut off from the rest of the world. Not having technology means not being able to participate in the circulation of knowledge and information, it means being marginalized, being excluded from any development process. The availability of connection thus becomes the most obvious symbol of the flourishing of a new type of inequality whose perverse effects are less directly related to the problems of scarcity, poverty and human rights than in the past. Being connected means having more opportunities to satisfy one's needs and having greater access to intangible assets that only social relations can offer, but at what costs to society?

From consumer to prosumer in the era of access

Within post-industrial society, living standards have improved exponentially, leading to a progressive shift of individual utilitarian interests from the question of survival to the question of the consolidation of an ever-higher quality of life style. In this knowledge society, the propensity for accumulation, the exploitation of wage labor and class conflicts are progressively giving way to concerns related to the need to always be informed, to have access to knowledge, to the problem of alienation and the characteristics of flexibility of new social actors. By questioning the universalisms, the rationalisms and the western aesthetic and cultural canons inherited from the Enlightenment (Loytard 1984), smart people are giving importance that the human experiential dimension is linked to culture, languages, aesthetics, art and to the symbols. This aspect is producing important effects both for what concerns the social domain, and for what concerns the consumer culture.

Assuming the form of social integration, the social domain is perpetrated through "the abolition of individual boundaries between the individual self and the organization" in the corporate world (Kunda 2000: 115) and through seduction, manipulation, and the incorporation into the world of consumption (Schmitt and Simonson 1997).

In this society control is no longer exercised by the capitalists but by those who hold knowledge, technical skills and education. It is a category of subjects that is increasingly identified with the technocratic *élites* and white-collar workers. Before them, in opposition, there are all those who for various reasons manifest conservative ideas that reject change or are fighting for a change based on different values.

Consumption thus begins to be interpreted according to two contrasting and paradoxical attitudes: on the one hand, there are those who embrace a hedonistic orientation typical of the consumer society centered on satisfying the greatest number of needs in the most intense way possible (Touraine 2002: 390), and on the other hand, those who develop an awareness of the negative impact of human actions on the environment and aspire to remedy waste by becoming *prosumers*.

The term *prosumer* – which is a combination of the terms producer and consumer – was first introduced in 1980 by Alvin Toffler[6] to indicate those individuals who choose to produce their own goods and services independently in order to sustain their consumption. Having quickly fallen into disuse, the term has experienced a new return to topicality during the development of the digital age, and in particular during the shift from Web 1.0 to Web 2.0, that is to say since users of the internet have ceased to be mere consumers of information and services to become content producers (Wikipedia), authors of judgments (eBay, Amazon, Tripadvisor), fashion influencers (Instagram, Pinterest, Blogs) and pioneers in the development of new customs and traditions (Facebook, twitter, Reddit).

The phenomenon of prosumerism in the context of knowledge capitalism has produced interesting contributions, widening its domain to the ideas of 'value co-creation' (Humphreys and Grayson 2008; Prahalad and Ramaswamy 2004; Zwick, Bonsu and Darmody 2008), of pro-am[7] (Leadbetter and Miller 2004), of 'wikinomics' (Tapscott & Williams 2006), of do-it-yourself (Watson and Shove 2008) and of 'productive consumption' (Laughey 2010), just to mention some.

Much of this transformation took place through the network, the social networking websites and the will of the smart people to have a direct impact on the socio-economic change underway. This meant that the Tofflerian concept of *prosumer* was revised to indicate – especially in the architectural and engineering field – those who adopt *smart grid* [8] and recycling systems for essential goods such as air and water[9] to decrease their ecological footprint on the world (Gurashi, Iannuzzi and Sessa 2019; Sessa 2019). Indeed, they become self-producers and consumers of energy, as well as consumers and recyclers of air and water. The *prosumer* is therefore today a consumer who at the time of consumption is in turn a producer or contributes to the production of what he consumes in goods and services. In this perspective the *prosumer* is the one who lives by equipping himself with technologies that allow him to reduce his negative impact on the environment in which he lives. He no longer identifies with the classic passive consumer but wishes to engage in creative acts of affirming his personality, through active intervention in all phases of the production process (Degli Esposti 2015). The affirmation of the personality also passes from the affirmation of the awareness of the human impact on the environment and on his will to actively contribute to change the contingent situation. A will that now remains anchored to the individual dimension and that needs the development of a 'smart politics' in order to be adequately declined (de Ghantuz Cubbe 2019) in its political, administrative and community dimension.

Smart choices: attention to environmental sustainability and ethical consumerism

The attention to ethical and environmental issues already highlighted in relation to the emergence of prosumerism, represents the logical foundation on which smart people have begun to build also a 'careful consumption' that balances hedonistic tendencies to satisfy needs with altruistic motivations deriving from the awareness and responsibility that individual and collective choices can have on the ecosystem. This consumption trend considers various aspects including: the ethical composition of products, the financial choices made by companies, the role played by corporate social responsibility in their strategic vision, corporate social marketing, and the choice or not to promote ethical causes by certain companies (Kolter and Lee 2005).

It is also a form of consumption that favors choices based on social and community motivations, often economically not logical or irrational and inefficient for Paretian optimality. These are choices, based on ethical beliefs of a personal nature (Carrigan et al. 2004) and which struggle to be framed in traditional canons of marketing since they presuppose processes and psychological evaluations that escape the cold laws of mathematical approximation. For these reasons it is no longer possible to think of the 'average consumer'[10], but to a series of ever-changing consumption experiences that are formed based not only on the psychological conditions of the individual, but also and above all on the basis of all those values, norms, customs, symbols, behavior patterns – which are often very different from region to region of the world – which give meaning to objects. This is a meaning that is often recognized not only by the individual subject, but also by the entire social group to which he belongs (Auger et al. 2003).

In addition to what has been said so far, it should also be noted that consumers of ethical products are often committed to detecting the "people element of consumerism" (Strong 1996) and the exploitation of the worker (in many cases underage) in underdeveloped and developing countries (Shaw and Clarke 1999: 109). Being an ethical consumer therefore also means "buying products which are not harmful to the environment and society. This can be as simple as buying freerange eggs or as complex as boycotting goods produced by child labor" (Harper and

Makatouni 2002: 289); it means giving priority to ethics and combining the value dimension with the utilitarian one.

In the case of the Millennials and therefore also of the smart people, the issue of ethical and sustainable consumption means not only being able to make smart choices that are smart because they are in line with the value dimension, but also making choices that can be implemented through the technological tools that the development of cognitive capitalism has progressively offered to consumers in order to improve their consumption experience.

The entry of technology into consumers' lives, combined with an awareness of the importance of ethical consumption, has contributed to changing the Consumer Decision Making Process (CDMP) originally described by Cox et al. (1983) which consists of the recognition of the problem, the search for information, the comparison of alternatives, the purchase, and evaluation in the post-purchase phase.

1 *Recognition of the problem:* At this stage the consumer becomes aware of having a need to be satisfied and does so by comparing his situation with that of another or an ideal image (which may be suggested by advertising). Today this first step of consumption is simplified and certainly expanded by the advertisements that invade our homes through television, computers, tablets, smartphones, radios and so on, conditioning the perceptions of ourselves, of the surrounding environment and creating previously unknown needs.

2 *The search for information:* Once the problem has been identified, the consumer begins to look for information on how to satisfy his need. In the era of cognitive capitalism this means connecting to the Internet and searching on websites, blogs, forums and social networks for the most suitable way to satisfy the previously identified lack.

3 *Comparison of alternatives:* After collecting information on the most appropriate way to satisfy one's needs, the consumer begins to compare the various alternatives of goods and services at his disposal and moves towards those goods and services that have more appropriate attributes to its cultural baggage and its priorities. Like the previous ones, this phase is also enhanced by the willingness to do research on the web that allows information to be obtained. Consumers are thus enabled to benefit from a broader assortment of products and services to choose from than the possibilities offered by traditional channels.

4 *Purchase:* Having resolved to make the purchase and having examined their options, the consumer makes the purchase.

5 *Evaluation in the post-purchase phase:* The evaluation phase is of fundamental importance not only for the consumer but also for the manufacturer of the product or service purchased. Was the product able to fulfill the promises made through marketing campaigns? Was the product adequate, exceeding or disappointing the buyer's expectations? If the consumer expresses a positive judgment on digital platforms, the brand will be able to benefit from it to convince of the quality of its products also the future eligible candidates who will have to choose between two similar products. Also, in this case the Internet offers visibility and an otherwise impossible circulation of information.

The relevance of the Internet in the consumer world is also leading to a gradual shift from traditional mass consumption to the consumption of cultural services and products. "Our society is therefore experiencing a new phase of capitalism characterized by a continuous impulse to be connected in virtual networks in order to obtain services and to try new experiences (as sport programs, virtual life-coaching, virtual tourism and so on)" (Gurashi 2017: 513).

This trend has also determined the emergence of a new consumption model: the 'sharing economy', otherwise called 'collaborative consumption'. It is a consumption model based on sharing and satisfying individual and collective needs through the sharing of goods, services, knowledge and experiences through social networks. This type of consumption is no longer based on the possession of goods as a determining element of the fluctuation of supply and demand on the market, but on the exchange and access to the goods and services that are desired without necessarily possessing them. This new economic model has among its objectives that of reducing the impact of the consumer society on the environment. Through the sharing economy it is indeed possible to be able to continue to consume products and services without this leading to an increase in the supply of consumer goods. The idea is that "the choice to access to services and products rather than to buy and own them reflects anti-consumerist inclination which demonstrates that it is not the possession, but the experience we made with these things that makes people happy. This inclination therefore is ecological in the sense that it drives us to optimize the use of assets we have at our disposition and to contain the excessive consumerism typical of 'Hyper-capitalism' (Gurashi 2017: 514). As social actors par excellence, smart people are contributing to radically change the meanings socially attributed to concepts such as 'property', 'demand and supply' of goods and 'market' by deconstructing and modeling them according to the new needs created by the phenomena of technological development and globalization, however, it remains to be ascertained whether these changes are able to have a positive effect on the reduction of the human footprint on the ecosystem.

Conclusion

Despite the determination to change the behavior and logic of the traditional market concept through a fast and new industrial revolution and consumption, some paradoxes stand out on the horizon. First, the contradiction between the aspiration to a more sustainable, fair and territorial capitalism and the progressive materialization of processes that reinforce the deregulation of markets and the globalization of needs and the deterritorialization of consumption emerges strongly. The diffusion of technologies and the circulation of knowledge that should have led to an increase in the well-being of workers and an improvement in lifestyles are instead causing an increase in company expectations of efficiency, speed and collaboration of the worker, but also to even control the biological functions of the employee through the use of smartwatch, video cameras, video analytics for facial recognition. A paradox is thus created between the aspiration to a condition of greater individual freedom and a contingent situation of increased forms of control and social domination. "Through the new technologies [...] the economy is invading every aspect of the lives of the human beings, commodifying aspects that once were not subordinated to the economic laws" (Gurashi 2017: 515) such as leisure and spare time.

Another obvious inconsistency in the eyes of sociologists is the contradiction between the aspiration to relocate the individual at the center of the political, economic and social system through the exaltation of his personality and the consequence of the technological advancement that is progressively putting the technology at the heart of the social system. Imagining a future without digital technologies and networks is now impossible for anyone and the digital divide represents a weapon to exercise new forms of domination.

Although smart people aspire to be not only promoters but also creators of sustainable development, the spread of the sharing economy is also producing unsustainable dynamics when it is favoring the deregulation of consumption and markets, which very often escape the real economy. This has the direct effect of making the financial and intangible economy grow at the expense of the real market.

Furthermore, while adopting *smart grid* or water recycling technologies, and therefore investing in technology capable of reducing consumption, smart people often end up with not smart attitudes by adopting unsustainable lifestyles that respond to a hedonistic trend of satisfaction of contingent needs able to reduce the benefits brought by the use of technological innovation.

Discussing these contradictions and identifying resolution strategies is very important to produce a further advance, this time not a technological one but a social one, able to diminish the ecological footprint of our society.

Notes

1 A value chain is a sequence of activities required to make a product or provide a service. The idea of the value chain becomes useful for analytical purposes once we include some features to the production process. "First, the activities are often carried out in different parts of the world, hence the term *global value chain*; second, some activities add more value and are more lucrative than others [...] third, some actors in the chain have power over the others" (Schmitz 2005: 4).

2 While classical capitalism found its foundation in the generation and circulation of considerable amounts of capital, in cognitive capitalism, the essential characteristic of post-modernity lies in the social valorization of the immaterial capital typical of the knowledge economy. For this reason, those able to hold scientific knowledge are able to hold power (Bell 1976).

3 All social researchers agreed that this generation should be included in a precise period from the mid 1980s to the early 2000s. However, it should be noted that it is not possible to identify a clear correspondence between the years that each scholar has considered as the periods of beginning and end of this generation. For Foot and Stoffman (1998) it extends between 1980 and 1995; for Howe and Strauss (2000) between 1982 and 1999 and for Twenge (2010) starting from 1982 onwards.

4 Considered as 'space of flows'.

5 Granovetter (1973) gives the example of the circulation of gossip that occurs faster and more pervasively among acquaintances and in a more slow and fragmentary way between family and close friends.

6 It should be noted that the concept of prosumerism was first introduced by McLuhan and Nevitt in their book *Take today. The executive as dropout* (1972) in relation to the production and consumption of electricity.

7 It is synonymous of semi-professional and indicates the collaboration or competition between professionals and amateurs in the sports and scientific field.

8 *Smart grids* represent a network composed of many other smaller networks, coordinated with each other, that connect producers and consumers. These mini networks are linked together so that they can exchange information. In this way the system can manage peak energy demand with maximum efficiency. In practice, this allows to avoid electricity interruptions and to reduce the load where necessary. By allowing efficient distribution of energy entering the network, *smart grids* manage to reduce the consumption of fossil fuels such as oil and coal.

9 Rainwater and gray water recovery systems for irrigation in agriculture, use of water as a process in a business environment, or for washing streets in urban centers and for feeding heating systems in the civil field.

10 "The Court of Justice defined an average consumer as someone who is reasonably well informed and reasonably observant and circumspect" (Cortés 2018: 6).

References

Antonini, E. (ed.) (2006) *Testimonianze sul capitalismo.* Roma: Bulzoni Editore.

Auger, P., Burke, P., Devinney, T. M. and Louviere, J. J. (2003) "What Will Consumers Pay for Social Product Features?" *Journal of Business Ethics* 42 (3): 281–304.

Bell, D. (1976) *The coming of post-industrial society. A venture in social forecasting.* London: Penguin Books.

Bell, D. (1991) *The Winding Passage: Sociological Essays and Journeys.* New Brunswick, London: Transaction Publishers.

Botteri, T. and Cremonesi, G. (2017) *Smart Working e Smart workers. Guida per gestire e valorizzare i nuovi nomadi.* Milano: FrancoAngeli.

Carrigan, M., Szmigin, I. and Wright, J. (2004) "Shopping for a better world? An interpretive study of the potential for ethical consumption within the older market." *Journal of Consumer Marketing*, 21(6): 401–417. doi:10.1108/07363760410558672.

Castells, M. (2001) *The Internet Galaxy.* Oxford: Oxford University Press.

Castells, M. (2009) *Communication power.* Oxford: Oxford University Press.

Cortés, P. (2018) *The Law of Consumer Redress in an Evolving Digital Market. Upgrading from Alternative to Online Dispute Resolution.* Cambridge: Cambridge University Press.

Cox, A., Granbois, Dh. and Summers, J. (1983) "Planning, search, certainty and satisfaction among durables buyers: a longitudinal study." *Advances in Consumer Research*, X: 394–399. Thirteenth Annual Conference. San Francisco. Association for Consumer Research.

de Ghantuz Cubbe, G. (2019) "Smart politics. The political dimension of the 'smartness'." In Iannone, R. and Gurashi, R., with Iannuzzi, I., de Ghantuz Cubbe, G. and Sessa, M., *Smart Society. A Sociological Perspective on Smart Living.* Abingdon: Routledge.

Degli Esposti, P. (2015) *Essere prosumer nella società digitale: produzione e consumo tra atomi e bit.* Milano: FrancoAngeli.

Foot, D. K. and Stoffman, D. (1998). *Boom, Bust and Echo 2000: Profiting from the Democratic Shift in the New Millennium.* Toronto: Macfarlane, Walter & Ross.

Granovetter, M. S. (1973) "The Strength of Weak Ties." *The American Journal of Sociology*, 78(6): 1360–1380.

Granovetter, M. S. (1983) "The Strength of Weak Ties: A Network Theory Revisited." *Sociological Theory*, 1: 201–233.

Gurashi, R. (2017) "The Sharing Economy at the Crossroads. A Conflict Between Social Values and Market Mechanisms." *European Journal of Sustainable Development*, 6(4): 511–516. doi:10.14207/ejsd.2017.v6n4p511.

GurashiR., IannuzziI. and SessaM. (2019) "From theory to empiricism: the challenge to the future of the Sapienza University at Solar Decathlon Middle East." In Iannone, R. and Gurashi, R., with Iannuzzi, I., de Ghantuz Cubbe, G. and Sessa, M., *Smart Society. A Sociological Perspective on Smart Living.* Abingdon: Routledge.

Harper, G. C. and Makatouni, A. (2002) "Consumer Perception of Organic Food Production and Farm Animal Welfare." *British Food Journal*, 104(3): 287–299. doi:10.1108/00070700210425723.

Harvey, D. (1990) *The Conditions of Postmodernity. An Enquiry into the Origins of Cultural Change.* Oxford: Blackwell.

Howe, N. and Strauss, W. (2000) *Millennials Rising: The Next Great Generation.* New York: Vintage.

Humphreys, A. and Grayson, K. (2008) "The intersecting roles of consumer and producer: A critical perspective on co-production, co-creation and prosumption." *Sociology Compass*, 2: 963–980. doi:10.1111/j.1751–9020.2008.00112.x.

Iannone, R. (2019) "Smart Society. The critical sense of a world strategy." In Iannone, R. and Gurashi, R., with Iannuzzi, I., de Ghantuz Cubbe, G. and Sessa, M., *Smart Society. A Sociological Perspective on Smart Living.* Abingdon: Routledge.

Iannuzzi, I. (2019) "Smart community: a new way of being together?" In Iannone, R. and Gurashi, R., with Iannuzzi, I., de Ghantuz Cubbe, G. and Sessa, M., *Smart Society. A Sociological Perspective on Smart Living.* Abingdon: Routledge.

Kagermann, H., Lukas, W. and Wahlster, W. (2011) "Industrie 4.0: Mit dem Internet der Dinge auf dem Weg zur 4. industriellen Revolution." *Technik & Gesellschaft*, 13. https://www.vdi-nachrichten.com/Technik-Gesellschaft/Industrie-40-Mit-Internet-Dinge-Weg-4-industriellen-Revolution (June 28, 2018).

Kamensky, E. (2017) "Society. Personality. Technologies: Social Paradoxes of Industry 4.0." *Economic Annals, XXI* 164(3–4), 9–13. doi:10.21003/ea.V164–102.

Kotler, P. and Lee, N. (2005) *Corporate social responsibility. Doing the most good for your company and your cause.* Hoboken: Wiley.

Kraut, R., Patterson, M., Lundmark, V., Kiesler, S., Mukopadhyay, T. and Scherlis, W. (1998) "Internet paradox: a social technology that reduces social involvement and psychological well-being." *American Psychologist*, 53: 1017–1031.

Kuhn, T. S. (1962) *The structure of scientific revolutions.* Chicago: University of Chicago Press.

Kunda, G. (2000) *L'ingegneria della cultura. Controllo, appartenenza e impegno in un'impresa ad alta tecnologia.* Torino: Comunità.

Latour, B. (1987) *Science in action: How to follow scientists and engineers through society.* Cambridge: Harvard University Press.

Laughey, D. (2010) "User authority through mediated interaction: A case of eBay-in-use." *Journal of Consumer Culture*, 10(1): 105–128. doi:10.1177/1469540509354137.

Leadbetter, C. and Miller, P. (2004) *The Pro-Am revolution: How enthusiasts are changing our economy and society.* London, England: Demos.

Loytard, J. F. (1984) *The Postmodern Condition: A Report on Knowledge.* Minneapolis: University of Minnesota Press.

McLuhan, M. and Nevitt, B. (1972) *Take today. The executive as dropout.* New York: Harcourt Brace Jovanovich.

Posada, J., Toro, C. and Barandiaran, I. (2015) "Visual Computing as a Key Enabling Technology for Industrie 4.0 and Industrial Internet." *IEEE Computer Graphics and Applications*, 35(2): 26–40. doi:10.1109/MCG.2015.45.

Prahalad, C. K. and Ramaswamy, V. (2004) "Co-creation experiences: The next practice in value creation." *Journal of Interactive Marketing*, 18(3): 5–14. doi:10.1002/dir.20015.

Schmitt, B. and Simonson, A. (1997) *Marketing Aesthetics: The Strategic Management of Brands, Identity, and Image.* New York: Free Press.

Schmitz, H. (2005) *Value Chain Analysis for Policy-Makers and Practitioners.* Ginevra: International Labor Office.

Sessa, M. (2019) "Home smart home. A sociology of living in the age of reflective materialism." In Iannone, R. and Gurashi, R., with Iannuzzi, I., de Ghantuz Cubbe, G. and Sessa, M., *Smart Society. A Sociological Perspective on Smart Living.* Abingdon: Routledge.

Shaw, D. and Clarke, I. (1999) "Belief Formation in Ethical Consumer Groups: An Exploratory Study." *Marketing Intelligence and Planning*, 17(2): 109–119. doi:10.1108/02634509910260968.

Soriano, J., Heitz, C., Hutter, H. P., Fernández, R., Hierro, J. J., Vogel, J., Edmonds, A. and Bohnert, T. M. (2013) "Internet of Services." In Bertin, E., Crespi, N. and Magedanz, T. (ed.) *Evolution of Telecommunication Services. The Convergence of Telecom and Internet: Technologies and Ecosystems.* Berlino: Springer-Verlag.

Strong, C. (1996) "Features Contributing to the Growth of Ethical Consumerism - A Preliminary Investigation." *Marketing Intelligence and Planning*, 14(5): 5–13. doi:10.1108/02634509610127518.

Tapscott, D. and Williams, A. D. (2006) *Wikinomics: How mass collaboration changes everything.* New York: Portfolio.

Toffler, A. (1980) *The third wave.* New York: William Morrow.

Touraine, A. (2002) "From understanding society to discovering the subject." *Anthropological Theory*, 2: 387–398. doi:10.1177/14634996020020041001.

Touraine, A. (2013) *La fin des sociétés.* Parigi: Éditions du Seuil.

Twenge, J. M. (2010) "A review of the empirical evidence on generational differences in work attitudes." *Journal of Business Psychology*, 25(2): 201–210. doi:10.1007/s10869–10010–9165–9166.

Twenge, J. M. (2017) *Gen: Why Today's Super-Connected Kids Are Growing Up Less Rebellious, More Tolerant, Less Happy – and Completely Unprepared for Adulthood – and What That Means for the Rest*. New York: Atria Books.

Watson, M. and Shove, E. (2008) "Product, Competence, Project and Practice." *Journal of Consumer Culture*, 8(1): 69–89. doi:10.1177/1469540507085726.

Zwick, D., Bonsu, S. K. and Darmody, A. (2008) "Putting consumers to work: 'Co-creation' and new marketing governmentality." *Journal of Consumer Culture*, 8(2): 163–196. doi:10.1177/1469540508090089.

4 Smart community

A new way of being together?

Ilaria Iannuzzi

Introduction to the smart community

The adjective 'smart' seems to connote more and more frequently any area of life. Even the size of the community currently appears to be increasingly called upon to become more and more intensely *smart*. But what does being a 'smart community' mean? When can the community be truly smart? What dimensions and processes does the smartness allude to? These are the questions that constitute the key questions of this chapter.

The starting point of reflection, in this sense, can only be constituted by the analysis of the conceptual category of the *community*, from which the conceptualization and the phenomenon of the *smart community* should start and to which they should probably be connected. The conditional is here mandatory, since the *smart community-community* relationship, as it will be possible to highlight, cannot be taken for granted.

The existing literature on the subject does not provide a unique definition of a *smart community*. This indefiniteness, if on the one hand does not help to clearly delineate what must be understood with the captioned expression and does not help, therefore, to limit the field of action, on the other hand already represents an important datum in itself, since it highlights how, not only is it a concept and a phenomenon in the making, but also how it lends itself to be shaped according to the aspect that we intend to emphasize the most, thus assuming flexible forms that are not rigidly pre-established.

Trying to summarize the multiple definitions of the *smart community*, one could use the following conceptualization, from which to start: a community can be said to be smart when it is able to act employing *new technologies in an active way with the aim of improving the quality of life of its inhabitants, mediating between the needs of citizens, institutions and businesses.* In this perspective, all those involved form real alliances "to work together on the development of technology applications aimed at transforming the community itself in a perspective of widespread well-being and sustainability" (Rizzi 2013: 103).

The technological element plays a central role in determining the smartness of the community. No doubt it represents a novelty within the classic category of community, but, on closer inspection, as will be emphasized below, it cannot be said to be the only factor capable of making the community in question smart.

Is *smart community* smart, compared to the classical community, exclusively because it has within itself the use of previously non-existent technologies? Following this hypothesis, the *smart community* would be nothing more than the simple and natural evolution of the classical concept of community: a community, with all the typical sociological characteristics of this conceptual category, in which the new technologies are massively used. What would therefore differentiate the *community* from the *smart community* would be the presence of ICTs, in the first case impossible to be used since they are materially non-existent. It would therefore be a 'historical' difference that focuses on a purely contingent element – a certain technological progress – leaving the meaning of the *community* category unchanged.

If things were like this, all in all they would be easy to interpret and would not leave much room for any exegetical or etiological doubts. On the other hand, however, the hypothesis presented does not seem to fully respond to the dynamics of the *smart community* phenomenon and it is not capable of exhausting the theoretical and practical complexity of this new way of thinking and living the community, if a new way it is.

Putting the question in these terms, it seems appropriate to go back to the meaning of the concept of *community*, to be able to effectively understand if and in what terms we can speak of the *smart community* as a new social phenomenon.

The concept of *community* continues, even today, to challenge any precise and rigorous definition and, as is known, there is a lot of literature on the subject (Bagnasco 1999; Bauman 2004, 2013; Pizzorno 1960; Tönnies 1979; Weber 1922). Indeed, it is often used to outline "social units" of various kinds, from family groups, villages, and neighborhood networks up to ethnic groups, national groups and international organizations (Shore 1997: 107). In sociological terms, the community is generally referred to as a group of people "who interact in the framework of shared institutions and who have a sense of common interdependence and a sense of belonging" (Shore 1997: 107). Even if, in the general meaning, we tend to identify the community with a well-defined geographical area, the spatial element represented by the territory does not seem to be an indispensable factor for the existence and life of the community itself.

This is well exemplified by contemporary *web communities* or *transnational communities*, which cross the border category by highlighting how the lack of a very precise spatial reference now does not constitute an obstacle for their development.

In addition, the simple coexistence and interaction of groups of people within a given territory is not a guarantee of a community. In order for the latter to develop, it is not enough for it to identify itself with a specific *structure,* but it must be sustained in a *state of mind*: "the feeling of being a community" (Shore 1997: 107). This feeling of belonging, of identity and identification with a whole, a greater totality of individuals, means that, within the community and towards the non-members of the community in question, "the values, the norms, the customs [and] the interests of the community" are, "more or less consciously", placed before personal interests or those of their own sub-group (Gallino 2006: 266).

The holistic character of the community is found in its being a "human whole" (Redfield 1965), "whose members live for and thanks to it" (Gossiaux 2006: 264). If on the one hand, therefore, the institutional dimension is underlined – the community as a structure – on the other, it is the *quality of the relationships* between the members of the community that seems to constitute its essence (Gossiaux 2006: 264).

Essential elements for the existence of any community would therefore be the sense of belonging to "a positively evaluated socio-cultural entity" and the experimentation of social relations capable of involving the person. Such elements in themselves generate solidarity (Gallino 2006: 266–267).[1] Note that this does not imply the automatic non-existence of conflicting moments within the considered community, nor the abolition of any form of domination or power within it (Gallino 2006; Mongardini 1993). Considering the community in symbolic terms means, therefore, thinking of it not so much as a "concrete collectivity", but rather as "a particular state that every community can temporarily assume" (Gallino 2006) and which changes from collectivity to community, depending on the precise social conditions they present.

As we can see, all this makes the attempt to apply the same definition of community, with the same content, to any social unit without distinction extremely difficult. For this reason, the concept of *smart community* also suffers from this problematic, provoking doubts about the applicability of the same interpretation of the adjective *smart* in any context (Iannone 2019). In other words, even the smartness required of community seems to depend on the specific characteristics of the considered community, from which derives the variability of the content of meaning with which the adjective *smart* is, from time to time, filled and, consequently, the uniqueness of this intelligence in relation to each community.

The community seems to emerge where social interaction is connoted according to certain "distinctive qualities" and therefore emerges as "a way of being of social relations" (Spreafico 2011: 144) capable of producing solidarity, trust and equality. In other words, there is community only when the social bond is generated (Spreafico 2011). The *smart community* is called on, in this sense, to produce a social bond, but if it has been necessary to coin a new expression, this aspect obviously does not seem to be enough to make it a *smart* social unit.

Then, what is the meaning of the invoked smartness? And how does the *smart community* fit with the classic categorization that sees the community as opposed to society?

The smart community in community-society dichotomy

To understand the potential and limits of the conceptual category of the *smart community*, it seems appropriate to focus the attention on the opposition *Gemeinschaft-Gesellschaft* by Tönnies. As we know, the *community*, according to Tönnies, represents a natural organism – it is "real and organic life" (Tönnies 1979: 45) – within which a common will and the collective interests predominate. Its members are limitedly individualized and a "global and spontaneous" solidarity reigns in it (Gallino 2006: 268). The moral is made up of religious beliefs and the subjects are inspired by customs.

As we know, *society*, on the other hand, represents an "ideal and mechanical formation" (Tönnies 1979: 45), an "artificial and conventional construction" (Bagnasco et al. 1997: 35) which is composed of individuals who are separated from each other, since each of them aims at satisfying one's own personal interest. Thus, society plays a role of a mere guarantee of the fulfillment, by individual contractors, of the obligations that they assume mutually (Bagnasco et al. 1997). In it, therefore, individual interests prevail, and individualization reaches its peak.

If in the community men live side by side with each other, being "essentially linked", in society they live "essentially separated" and remain in such a situation of separation "despite all ties". In the community, on the other hand, they "remain linked despite all the separations" (Tönnies 1979: 83).[2]

The Tönnesian reflection on the subject is therefore radically opposed, as it does not let forms of solidarity or social integration affirm themselves within society. From this point of view, unlike the Durkheimian interpretation, it is not the typical solidarity of complex society, but the community that is considered organic. If the transition from community to society inevitably entails, according to Tönnies, the loss of solidarity, this does not happen following the Durkheimian exposition, which supports not the elimination of solidarity, but its transformation, its change of face – from a dimension of interchange-ability to one of interdependence – in the transition from the pre-modern to the modern society (Rutigliano 2001). Therefore, it is not the advent of the complex society that does not contribute, but rather the affirmation of pathologies connected to the anomie, on the one hand, and to coercion, on the other, i.e. when the division of labor assumes the anomic or coercive form (Rutigliano 2001) that con-tributes to make this element fail.

It should be noted that in the Durkheimian perspective the commu-nity coincides with that "intermediate organ between the private life of the family and the public life of the State" (Esposito 2011: 10) and how, therefore, beyond the various interpretations, the same purpose assigned to the Community element, namely ensuring social cohesion, recurs in both cases. The nature of community becomes of "economic-incarnational" nature (Esposito 2011: 10), to say, it aims at embodying an image in which everyone can "recognize, identify and believe" as "part of a whole" (Esposito 2011: 10), as a unit.

The farsighted reflection of Schleiermacher who conceives the com-munity no longer as a "supra-individual entity" through which the indi-vidual can go beyond his limited human condition, realizing "goals that transcend his own strength and the duration itself of his existence" (Gal-lino 2006: 268) – as was typically the case in romantic thought –, but as a "particular bond between individuals", i.e. a social relationship, which finds its foundation in a common external purpose (Gallino 2006: 268) appears interesting for the purpose of the current investigation. Unlike mere reciprocity, the community understood in this way is again opposed to society, defined as a form of "social purpose without purpose" (Schleiermacher 1799).

Similarly, the Weberian differentiation between the "unifying" – *Ver-gemeinschaftung* – and the "association" – *Vergesellschaftung* – (Weber 1922) also depends on the subject's will and his orientation to action: a community is created whenever the orientation towards action is based "on the mutual belonging subjectively felt by the members", while society manifests itself when this orientation is erected "on rationally motivated interests" (Gallino 2006: 269).

Considering these reflections, how can the *smart community* be under-stood? It is natural to ask whether it, as a community, refers to the decli-nations of communities highlighted above or if it deviates, and possibly in what terms, from the classical conception of the community that con-siders it necessarily in contrast with society. These questions emerge with even greater vigor if we consider the high degree of specialization and individualization that characterizes the contemporary social order. So, which of the two poles should the *smart community* be located in?

For the analysis to be carried out, we should point out that in the Anglo-Saxon meaning, the term *community* generally assumes the meaning of "local community, town or village or suburb" (Gallino 2006: 269). In this sense, the community thus understood is nothing more than the most suitable category to identify the current communities within smaller territories of the city. Community is here synonymous of urban agglomeration of small / medium dimensions, that is a whole – whether a group, an aggregate or a system – of subjects that live in a particular place (Strassoldo 1987: 485). The community is therefore defined as 'ecological', according to a typically spatial meaning that puts the territorial dimension of reference in the foreground (Strassoldo 1987). Often, however, we face the presence of communities of people that go far beyond the city limits. It is, in particular, the case of the *smart community*, which is frequently understood as a synonym of a *widespread* city, thus crossing the dimensions of the urban agglomeration of medium and large size (Niger 2012). What can we do, then? (Strassoldo 1987: 485).

The term *city*, generates considerable confusion – because if the only criterion that distinguishes the *community* from the *city* is represented by the territorial greatness, identifying the distinction between the two can, in some cases, become extremely complex – and also this meaning of the concept of community does not seem to totally and fully respond to the degree of complexity that this phenomenon possesses and does not appear to be able to grasp its many nuances of meaning. This for two reasons.

The first reason is constituted by the evidence that in all the fields of the humanistic disciplines meanings of the concept of community, that even if in different ways, recognize centrality to the elements of the identification and the belonging of the members towards of the community, have been elaborated over the centuries. As we have seen, the ecological community in itself is not a guarantee of the existence of a *feeling* of community (Strassoldo 1987), it does not generate *social belonging*, but *participation* (Spreafico 2011). Such an interpretation, focusing not on the symbolic meaning of the community, but on its geographical territory, appears, therefore, to be reductive compared to the sense that over time the concept of community intended to express.

The second reason is linked to the observation that the spatial element, the proximity-distance relationship, becomes less and less decisive for the functioning of the current social order, as evidenced by the decline in the role exercised by local aggregations – for example, neighborhood, area, village – in the daily life of the subjects (Strassoldo 1987). It should be noted, in this sense, that, in recent times, we have forcefully been appealing to the return of importance of these levels of grouping, which sometimes results in the desire for the advent of new forms of real communalisms. Nevertheless, it is evident that currently the concept of community is difficult to describe exclusively according to an interpretation bound to territorial contexts and geographical proximity. Online communities exemplify this new reality (Paini 2012).

By placing the discourse within the outlined perspective, the *smart community* appears not only as a merely geographic aggregate, but as a social unit that includes its own culture, its own identity, a specific social interaction extended internally and towards the external (Winthrop 1991).

Placing the *smart community* totally at the pole of the community could seem to be the operation most responsive to reality, since, in both cases, it is a matter of community. However, if we understand the community as an expression of opposition to society, we note that this approximation involves some problems.

First, this would mean understanding the *smart community* as a nostalgic need to recover pre-modern forms of social life and would therefore imply the attribution of a positive value to the community as it is preferable to the way society operates. The risks connected to the affirmation of real community ideologies appear, in this case, remarkable.

In this direction, moreover, if on the one hand the characteristics that are deemed lost with the transition to modernity are enhanced – that is, using a parsonsian terminology, interaction based on affectivity and orientation in view of the community against affective neutrality and orientation in view of the ego (Parsons 1951) –, on the other, just as energetically, the more 'inconvenient' aspects that life in the community, according to the classical meaning, entails are forgotten. Often identified as the realm of the oppression of the individual, through ties that are considered laces, which do not leave the subject free, the community in opposition to society has frequently been considered as limiting the freedom of the individual. With modernity and the advent of society, it was hoped, therefore, that the traditional community based on class bonds and on the role of authority would be replaced with a new social unit building on the model of man in the abstract, a subject perceived exclusively "as pure freedom and desire for possession" (Barcellona 1990: 75).

Considering the *smart community* as an expression of the desire to recover these modes of social functioning would imply, therefore, the negation of all that is 'positive' and 'desirable', for example in terms of freedom, which society represents, such as exemplified by the dynamics company founded on the Status-company based on the Contract (Maine 1861).

On the other hand, the total juxtaposition of the *smart community* to society, with all its typical characteristics, does not show itself as the option that best meets the real. It seems plausible, in fact, that a *smart community* will not be considered very *smart* and very *community* if based on individualistic criteria of satisfaction of personal interests and if based exclusively on an instrumental rationality.

The concrete challenge that the *smart community* therefore seems to face is not so much its identification with the pole of the community or with that of society, but rather its insertion along the continuum represented by the two poles. The *smart community*, as has been pointed out, cannot coincide with the classic ways of conceptualizing the community, nor with those of representing society. The objectives that it sets itself – including improvement of the quality of life, strategic vision for the future, active citizenship, sustainable mobility, cultural development, innovation – cannot avoid dealing with values, behavioral models, cultures, institutions, relationships of interaction and power – think of the role of politics and, in particular, of *smart politics* (de Ghantuz Cubbe 2019) – typical of the current social structure.

If we assume the community as a way of acting, as a "way of being of social relations" (Spreafico 2011: 144) aimed at producing social ties, it becomes feasible even within the typical working methods of contemporary society and not necessarily in opposition to the latter. The typicality of the *smart community* lies, therefore, in the ability to combine the typical elements of the community – solidarity, mutual belonging, social cohesion – with those proper to society. In this sense, the *smart community* does not represent the denial of instrumental rationality or any other element proper to modern society, but the affirmation of the possibility of creating "active relational connections" (Paini 2012) able to guarantee social objectives such as inclusiveness, sustainability, innovation and social cohesion *within* the current social structure and without running away from it, ending in pre-modern community forms.

It is therefore not the community as such that is here called on to be reconstituted at present, but the *community relationship* within today's society (Spreafico 2011). Fertile ground, in this sense, is represented by "everything that cannot be rationalized", such as the dimension of values or that of affects (Spreafico 2011: 150), dimensions that can never be completely excluded from social life and which are always to be found within it, exemplified by the mechanical solidarity *enclaves* of Durkheimian memory (Rutigliano 2001).

The *smart* element of the community is therefore embodied in an *approach* that is closely connected to the importance of human, relational and social capital. It is for this reason that only the technological factor proves insufficient to generate smartness (Manfredi 2015). Therefore, in the combined combination of community and society and not in the oppositional relationship, the smartness of the *smart community* is manifested.

Smart community and smart city: features and differences

The conceptual category of the *smart community* is often used as a synonym for *smart city*. In the public debate, but frequently also within the specialist literature on issues related to *smartness*, the terms *smart community* and *smart city* are used interchangeably, to indicate realities if not identical, at least easily comparable for their characteristics. But is it irrelevant, from the semantic point of view, to consider *cities* and *communities* as synonyms?

It was pointed out that the concept of 'ecological community' refers to the territorial dimension as a key to this specific way of understanding the community. The link with the territory appears, therefore, as a recurrent factor, but, nevertheless, insufficient to create community. It is through this perspective that cities and communities, in their *smart* sense, can be more easily confused. As mentioned earlier, it becomes extremely complicated to define what a city is and what community is if the spatial criterion is used as the sole principle of difference between the two. The biggest confusion, in any case, is generated where the *smart community* is considered a sort of evolution of the *smart city*. In fact, the need for a transition from the *smart city* to the *smart community* is often stated, that is, the need for the city to become more and more a community.

The *smart city* is a city, therefore a well-defined territorial context, which has multiple objectives and whose smartness is rooted in the ability to guarantee these objectives in a systematic and integrated manner. Smart urban space is characterized by its ability to face emerging social challenges, in terms of sustainability, security, education, cultural development, innovation and inclusion. In the *smart* perspective, these objectives are achievable using ICT, so the *smart* city is necessarily a technologically advanced city and whose technologies make it possible to raise the level of quality of life of those who live there. This does not mean, however, that the *smart city* is a totally 'technicized' city, i.e. a city with complex technological systems that overshadow human will and relegate the inhabitant to the figure of a mere final user (Olmedo Moreno and Delgado López 2015; Colado et al. 2014). This, in fact, would be in contrast with what the protagonists of the current change of action in the urban area propose, the so-called *smart people* (Gurashi 2019), who choose and decide on the basis of objectives that are not initially considered central, such as, for example, environmental sustainability hope.

In this direction, the city, in addition to being *smart*, is also *sentient* (Shepard 2011), that is able to remember and anticipate (Niger 2012), therefore, to exercise typically subjective abilities. This highlights an aspect that the *smart* adjective alone fails to sufficiently highlight, that is, that the complex data generated in the cities does not derive automatically or necessarily from an objective process, but presupposes, upstream, a subjectivity. Therefore, decisions, wills and the ability to think constitute the premises for the development of the smartness of the city. Thus, in line with what was asserted in the field of *Science and Technology Studies,* this contributes to remembering that there are no 'objective' technologies – since the development of a given technology itself presupposes a choice in this sense – and 'neutral' – as the complexity of the social is transfused within algorithms that decide and make choices and the same power relations are encapsulated in software codes able to act (Niger 2012).

Where the importance of the social context of reference in influencing elements such as scientific research and technological innovation is forgotten and where it is ignored how do these, in turn, influence the social order (Kuhn 1962; Latour 1987; Volti 2001); these factors are not framed "within an overall and systemic vision of the city and its future" (Mochi Sismondi 2012), but fragments remain, of the "pieces of a mosaic which drawing cannot be read" (Mochi Sismondi 2012).

Since during the concrete application of *smart city* urban models, the technological component often prevails, declined, in particular, according to an efficient approach that places the achievement of efficiency as such as its main purpose, all this has led, in the collective imagination, to the *smart city = ICT* equivalence. For this reason, we currently tend to speak not only of *smart cities,* but of *human smart cities* (Olmedo Moreno and Delgado 2015; Rizzo 2015; Meloni et al. 2017), to say a *smart city* not only because it is technologically gifted, but above all because it is capable of placing the human being and his needs at the center of his model.

In this direction, the *smart community,* expressing a certain type of relationality – based on the belief in shared institutions, on the sense of belonging and commonality of the subjects to a reality that goes beyond the single, on the union and unity between the parts – it is perceived as the necessary 'evolution' of a *smart city* which, presented purely from a technological and efficient point of view, has shown its weakness. From this point of view, the *smart community* would be nothing but the most advanced version of the *smart city*.

On closer inspection, however, reality appears to be different and the classically conceived conceptual categories of *community* and *society* reveal themselves to be enlightening, highlighting how a specific "relational state" that the community can take (Gallino 2006) – the community – cannot be compared to a place or a territorial context – the city.

Although these are two different concepts, they show a remarkable interconnection. Although, on the one hand, the territorial / spatial reference has, in fact, today diminished in importance due to the global globalization processes, on the other it continues to play a central role, as, in practice, it is on the land physically understood – on a specific geographic-territorial space – that numerous problems arise. Consider, for example, how the comparison and coexistence between different communities are played, often, within a given space. The city, in this direction, seems to represent the privileged space within which the processes in question are manifested.

Cities and *communities* therefore intertwine again, but their relationship cannot be taken for granted. It should be observed, for example, that in a city multiple communities can coexist or that only one community can correspond to it – where the community is understood as a specific collectivity, according to a meaning that, ultimately, assimilates the community to the social unity of the *group* (Mongardini 1993) –, but also how, within the city, there may be no community at all, if by this we mean a particular relational mode.

In light of these considerations, it seems appropriate to consider the *smart community* not so much as a development of the *smart city*, but rather as a possible component of the latter. The *smart community*, as an element founded on the increase of social capital through relational dynamics that move from the individual to the community (Meloni et al. 2017), can take place within the urban space – and, therefore, to be realized *into the smart city* and not *after* or *beyond* it – but it can also develop in places other than the city or, even, not take root anywhere.

While both refer to the current social challenges represented by sustainability, inclusion, innovation, technology, resilience, mobility and so on, *cities* and *communities* are two different concepts: the first is more closely tied to a geographical context, while the second is more flexible and not necessarily declinable spatially. So, *smartness* also takes on different meanings in both cases. If the city shows itself to be smart where it is able to guarantee the simultaneous and integrated satisfaction of all the needs mentioned above, the community becomes smart when the interaction between the parts that compose it generates knowledge, combining two fundamental elements: capital social and development (Manfredi 2015).

The *smartness* of the community, in this sense, is closely connected to the role of community social capital and to the capacity to develop permeable boundaries (Manfredi 2015), suitable for ensuring real inclusiveness without falling into the exclusive exclusivity typical of the classically conceived community. The real challenge to the smartness of the community is substantiated, therefore, in the activation of community relational modalities that not only affirm themselves internally, and no longer in opposition to society – taking into account its current characteristics and overcoming, therefore, a theoretical model marked by the dichotomy between affectivity and rationality as inevitably opposed modes of conduct – but which are also different with respect to the community conceived in antithesis to society, or modalities aimed at producing an inclusive social bond for communities that do not prove to be self-referential and closed on themselves.

The role of trust

An attempt has been made to highlight how the recurrent element, when speaking of the community, is constituted by the idea that a community dimension is developed whenever social ties present themselves "in solidarity and cooperation" (Spreafico 2011: 158), and act as a guarantee for social cohesion. In this sense, the community becomes that social subsystem aimed at safeguarding the integrative function of society. Using a parsonsian expression, the community is a "corporate community" (Parsons 1994) and represents the most important subsystem for the maintenance of society itself.

The processes linked to the growing individualism and economism of contemporary society, together with the process that led to the unraveling of those "intermediate entities of human aggregation" (Strassoldo 1987: 494) generate a reiterated need for community today, as a social unit capable of providing security: the individual, in his condition of atomicity, anomie and massification finds in the community dimension the resolution to the sense of lack of ties (Strassoldo 1987).[3]

We therefore risk falling into a communitarian ideology or a form of neocomunitarism (Etzioni 1996), which, by contrasting the market with society,[4] affirm the impossibility of bringing the first sphere back into the second. Having changed what needs to be changed, this is tantamount to disregarding the possibility that community-type relational modes can be made within the current social order and leads to the idea of community as a rescue anchor to which to cling in a sinking society, a particularistic community in which to shut oneself up and which, in turn, is closed to the outside.

In order to prevent the *smart community* from following the community's exclusive modes of interaction according to the classical meaning, the role played by the fiduciary element is revealed. If the community can be introverted, exclusive and defensive towards the outside, the *smart community*, to be such, can only be receptive and inclusive, inspired by behaviors dictated by generalized trust, in which, therefore, particularism not only does not prevent development, but rather encourages it.

The different types of trust with their different characteristics come into play. In addition to personal trust – aimed at the individual – and interpersonal trust – addressed to a collective actor – systemic trust becomes fundamental, that is to say trust towards society as a whole or by considering only some of its subsystems (Luhmann 2002). Personal trust, in fact, faces numerous difficulties in the presence of a highly complex and differentiated social system, such as the current one. Systemic trust, rising to a more impersonal and abstract level, intervenes in the attempt to reduce the existing social complexity (Bassi 2000). However, the process of extending systemic trust in today's society does not turn out to be easy to achieve, especially where the recipient of the trust expectation is a type of society – the *smart society* in the captioned case, whose features remain, in practice, difficult to identify. In this sense, when systemic confidence decreases, personal trust does not automatically decrease, but tends to grow. Keeping trust at the micro-level thus allows the macro-level to be protected from the disappearance of systemic trust (Luhmann 1989).

Smart reality, therefore, can hardly do without the presence of both personal and systemic trust. Confidence in *smart* dimensions, in fact, can be effectively nourished through the intensification of interpersonal trust that is established in the relationships between the subjects, through, therefore, the presence of *smart* relationships. Thus technological, organizational and commercial trust can be positively supported by personal trust (Sztompka 1996).

In this sense, the community cannot be qualified as *smart* if it does not recognize the driving role of the relational dynamics. These dynamics are characterized by faith in a trust which is not exclusive and limited to the members of the community, not only systemic – as emerges when referring, for example, to consumer confidence in technological products – but open to the outside, inclusive and enhancing interpersonal ties.

The virtuality typical of social relations of the contemporary era makes it difficult to establish forms of *smart community*. It is complicated by the process of globalization which, given the immense distance in time and space between the countries of the globe, requires a high degree of trust in order to pursue the expansion of social relations at a global level, through an enormous effort and an equally large investment (Pendenza 2000). The "socially positive dispositions" (Mutti 1998: 35) of the subjects are called to change in the "overcoming of the boundaries imposed by parental solidarity" (Mutti 1998: 19) and by the exclusivist solidarity typical of the classically understood community, moving towards the direction of the generalization of trust.

Conclusion: the future of smart community

In light of the reflections made, the conclusions to this contribution certainly do not intend to outline a decisive outcome on the *smart community*. These are, in most the cases, food for thought that are not conclusive elements, but a starting point for further future research that will tackle the topic from a sociological point of view. Today, the affirmation of the *smart community* encounters some difficulties, starting from the various types of obstacles that the very idea of community and its practical forms are called on to face today.

As we have tried to highlight, the s*mart community* does not act as a counterpart to society as it is known today, nor does it aim to reconstitute outdated forms of exclusive and closed communities. It tends to recover the relational modalities typical of the community context, but within the current social structure, not avoiding confining itself in a distinct and separate space. Moreover, these relational methods are based on the principles of openness, accessibility, cohesion and inclusiveness. The borders of this form of community are, therefore, always "flexible and permeable" (Manfredi 2015: 38).

In this sense, today's process of "deterritorialization of the social" (Spreafico 2011: 159), which leads to the gradual reduction in the importance of geographical proximity with regard to the affirmation and development of social ties must be highlighted. Consider, for example, how the contemporary dimension of the 'risk' unites spatially distant communities, up to talking about real *communities of risk*. The obsolete ideas of the "community of the human race" and of the "planetary community" are, therefore, replaced by social relations formed through continuous mechanisms of disaggregation and aggregation (Spreafico 2011: 159).

The dissemination of elements such as insecurity and uncertainty easily translates into the formation of communities whose members are linked only by a "commonality of conditions" (Pizzorno 1999: XXXIII). These communities seem to reflect the professional groups constituting the solidarity of defense of Durkheim's memory. That is, they are merely constituted by the fact that the subjects share the same condition and, through their union, intend to change the condition in question. Such solidarity, which finds its roots within material interests and relations – similar, therefore, to the class solidarity drawn by Marx – can only be far from the constitution of the *smart community*. The latter, on the other hand, cannot but deal with these problems.

Among the new forms of community, the so-called "aesthetic community" stands out (Maffesoli 2004). These are communities that are based on the sharing of the emotional element: unstable communities, not necessarily located in a territorial space, in most cases at the end. Communities that are based, therefore, on the "crossing of aesthetic codes, of image flows", on the importance "of fashion and lifestyles as producers of shared meaning" (Spreafico 2011: 161).

The *smart community*, whose conception is far from the proper sense of the aesthetic community, where it is understood as an opportunity to share ephemeral and non-design experiences, risks falling into the purely aesthetic mechanism, which places togetherness as the ultimate goal through empathic aggregation. With no reference whatsoever to common goals and values, "an experience of community sharing is evoked without any real community" (Spreafico 2011: 162) the common good is replaced by "aesthetic codes positively evaluated in a specific period" (Spreafico 2011: 161). The lasting social bond, which is at the base of the *smart community*, is so eroded, leaving room for the affirmation of the experiential criterion. The same fiduciary element loses its meaning in this perspective.

Here, then, are some of the problems that derive from a certain interpretation of the *smart community*. Its ability to assert itself in time and space and, therefore, its duration ultimately depends on the representation it is intended to make. Where the emotional and experiential element prevail, the *smart community* suffers an interpretation that distorts its constitution, making it a phenomenon destined, like all ephemeral phenomena, not to endure. The attention, therefore, should be given, in particular, to those interpretations that make the *smart community* a phenomenon dictated by fashion. In this case, it would have a short life, becoming a "piece of an unread and operationalized system integrated with the other parts" (Manfredi 2015: 24).

The originality of the *smart community*, therefore, is believed to be found in the relevance that it attributes to the relational element as an end that then becomes a tool: an end to be pursued, as an element capable of generating "active relational connections" (Paini 2012) and lasting social bonds. These elements, in turn, become fundamental tools and simplifiers to achieve *smart* goals, connected to inclusion, sustainability, technological innovation and so on.

The technological dimension, in this context, represents an essential element in order to make smart the ambit that it intends to transform. In this sense, one of the critical points is constituted, as we have tried to highlight, by all those situations in which technology is disconnected from the real needs of the subject and is detached from the relational dimension of society. Therefore, there remains the problem related to the possibility of putting into practice technologies that take into account the relationality and that are based on it, of generating real technologies applied to the objectives of the *smart society*: to inclusion, to sustainability, to all those aspects that must be based on a developed relational fabric in order to be inverted.

Alongside technology, an equally important role is now being played by the digitization process. In a critical perspective, a *smart* community that conceives, for example, reproduces forms of *digital divide*, undermining its inclusive capacity does not seem conceivable. Think of the risk of exclusion of disabled or elderly from the use of certain technologies, a problem that also emerges in the domestic sphere, for example, with the so-called *smart* home: a home that is completely domotic, but often incapable of meeting the basic needs of disabled users (Sessa 2019; Gurashi et al. 2019).

The technology not applied to inclusion therefore risks being transformed into the submission of "non-empowered individuals" by "empowered individuals" and to generate a real "globalization of exclusion" (Russo 2017). Such, therefore, are some of the challenges that the *smart community* is called on to face for its realization.

In it, the source of union between the members of the community is no longer understood as something that can be appropriated, but as a value that unites without generating separations and fractures from what is external to it. The *smart community*, as has been pointed out, overcomes the old community-society opposition. If, in fact, on the one hand, in the classically understood community the subjects are linked to each other, but still divided into many closed and self-referential communities and into society, on the other, all the subjects are separated from each other, in the *smart community* the subjects not only recover the reciprocal bond, but extend it: it is no longer substantiated in many opposing collectivities. The subjects, therefore, return to be linked to each other within inclusive and open social units, which greatly expand the social bond and are thus better able to achieve the objectives that define the very smartness of the community.

The future of the *smart community* and the possibility of it becoming that "spirit community" of Tönnesian memory (1979: 57) seems to play a role in this area. It seems, therefore, plausible to believe that only through these methods and only where the *smart community* is not reduced to the *connected community*, will it really represent a new way of being together and not simply a transient phenomenon.

Notes

1 Note that, among the different criteria that over time have been used to identify the nature of the community, the affective criterion has frequently been used, which, among the dimensions characterizing the community, stresses that relating to the need for existence, solidarity between the members of the community. Among the authors who have understood the community in this sense, see, specifically, Kanter (1972).

2 In society, says Tönnies, "no one will do anything for the other, no one will want to give something to the other, except in exchange for a service or a mutual donation that he considers at least equal to his" (1979: 83). It therefore seems that all forms of solidarity disappear.

3 As Bauman states, "The fact that the community is always present makes us feel secure. It is not something fluid, liquid. It never abandons us; whenever we need to refer to the place to which we belong, it is always there waiting for us and this gives us comfort" (2013: 32–33).

4 Consider, for example, Polanyi's reflection when he states that the market advances on the desert of society (1944).

References

Bagnasco, A., Barbagli, M. and Cavalli, A. (1997) *Corso di sociologia.* Bologna: il Mulino.

Bagnasco, A. (1999) *Tracce di comunità.* Bologna: il Mulino.

Barcellona, P. (1990) *Il ritorno del legame sociale.* Torino: Bollati Boringhieri.

Bassi, A. (2000) *Dono e fiducia. Le forme della solidarietà nelle società complesse.* Roma: Edizioni Lavoro.

Bauman, Z. (2004) *Voglia di comunità.* Roma-Bari: Laterza.

Bauman, Z. (2013) *Communitas. Uguali e diversi nella società liquida.* Milano: Aliberti.

Colado, S., Gutiérrez, A., Vives, C. J. and Valencia, E. (2014) *Smart city. Hacia la gestión inteligente.* Barcelona: Marcombo.

de Ghantuz Cubbe, G. (2019) "Smart politics. The political dimension of the 'smartness'." In Iannone, R., Gurashi, R., with Iannuzzi, I., de Ghantuz Cubbe, G. and Sessa, M., *Smart Society. A Sociological Perspective on Smart Living.* Abingdon: Routledge.

Esposito, M. (2011) *Oikonomia: una genealogia della comunità. Tönnies, Durkheim, Mauss.* Milano-Udine: Mimesis.

Etzioni, A. (1996) "The Responsive Community: A Communitarian Perspective." *American Sociological Review* 1, 1–11, doi:10.2307/2096403.

Gallino, L. (2006), *Dizionario di sociologia* (A–I), vol. I. Novara: Istituto Geografico De Agostini S.p.A.

Gossiaux, J.-F. (2006) "Community." In Bonte, P. and Izard, M., eds., *Dizionario di antropologia e etnologia.* Torino: Giulio Einaudi editore, 264–265.

Gurashi, R. (2019) "Smart people: the individual challenge of the fourth industrial revolution." In Iannone, R., Gurashi, R., with Iannuzzi, I., de Ghantuz Cubbe, G. and Sessa, M., *Smart Society. A Sociological Perspective on Smart Living.* Abingdon: Routledge.

Gurashi, R., Iannuzzi, I. and SessaM. (2019) "From theory to empiricism: the challenge to the future of the Sapienza University at Solar Decathlon Middle East." In Iannone, R., Gurashi, R., with Iannuzzi, I., de Ghantuz Cubbe, G. and Sessa, M., *Smart Society. A Sociological Perspective on Smart Living.* Abingdon: Routledge.

Iannone, R. (2019) "Smart Society. The critical sense of a world strategy." In Iannone, R., Gurashi, R., with Iannuzzi, I., de Ghantuz Cubbe, G. and Sessa, M., *Smart Society. A Sociological Perspective on Smart Living.* Abingdon: Routledge.

Kanter, R. M. (1972) *Commitment and Community: Communes and Utopias in Sociological Perspective.* Cambridge, Mass: Harvard University Press.

Kuhn, T. (1962) *The structure of scientific revolutions.* Chicago: University of Chicago Press.

Latour, B. (1987) *Science in action: How to follow scientists and engineers through society.* Cambridge, MA: Harvard University Press.

Luhmann, N. (1989) "Familiarità, confidare e fiducia: problemi e alternative." In Gambetta, D., ed., *Le strategie della fiducia. Indagini sulla razionalità della cooperazione.* Torino: Einaudi.

Luhmann, N. (2002) *La fiducia.* Bologna: il Mulino.

Maffesoli, M. (2004) *Il tempo delle tribù. Il declino dell'individualismo nelle società postmoderne.* Milano: Guerini e Associati.

Maine, H. S. M. (1861) *Ancient Law: Its Connection With The Early History Of Society And Its Relation To Modern Ideas.* London: John Murray.

Manfredi, F. (2015) *Smart Community. Comunità sostenibili e resilienti.* Bari: Cacucci editore.

Meloni, C., Tundo, A., Paoloni, G., Orsucci, F. and Cervini, F. (2017) "Dalla Smart City alla Smart Community." *Energia, ambiente e innovazione* 1, 40–45, doi:10.12910/EAI2017–2006.

Mochi Sismondi, C. (2012) *Non facciamo diventare la Smart City una moda vuota,* editoriale Forum PA. https://www.forumpa.it/citta-territori/non-fa cciamo-diventare-la-smart-city-una-moda-vuota/ (February 14, 2019).

Mongardini, C. (1993) *La conoscenza sociologica.* Genova: ECIG.

Mutti, A. (1998) *Capitale sociale e sviluppo. La fiducia come risorsa.* Bologna: il Mulino.

Niger, S. (2012) "La città del futuro: smart city, smart community, sentient city." http s://s3.amazonaws.com/PDS3/allegati/Niger_smart-city-articolo-1.pdf (February 14, 2019).

Olmedo Moreno, E. M. and Delgado López, A. (2015) "From Smart Cities to Smart Human Cities: Digital Inclusion in App's." *Revista Fuentes* 17, 41–65, doi:10.12795/revistafuentes.2015.i17.02.

Paini, G. (2012) "Cosa sono le Smart Communities." http://www.unithinktag.it/ it/resources/cosa-sono-le-smart-communities (February 14, 2019).

Parsons, T. (1951) *The Social System.* New York: The Free Press.

Parsons, T. (1994) *Comunità societaria e pluralismo. Le differenze etniche e religiose nel complesso della cittadinanza.* Milano: FrancoAngeli.

Pendenza, M. (2000) *Cooperazione, fiducia e capitale sociale. Elementi per una teoria del mutamento sociale.* Napoli: Liguori Editore.

Pizzorno, A. (1960) *Comunità e razionalizzazione. ricerca sociologica su un caso di sviluppo industriale.* Torino: Einaudi.

Pizzorno, A. (1999) *Introduzione a Durkheim É., La divisione del lavoro sociale.* Milano: Edizioni di Comunità.

Polanyi, K. (1944) *The Great Transformation.* New York: Farrar & Rinehart

Redfield, R. (1965) *The Little Community.* Chicago: The University of Chicago Press.

Rizzi, F. (2013) *Smart city, smart community, smart specialization per il management della sostenibilità.* Milano: FrancoAngeli.

Rizzo, F. (2015) "Design and Social Innovation for the Development of Human Smart Cities." *Nordes 2015: Design Ecologies* 6, 1–8.

Russo, F. (2017) "Presentazione. Miglioramento, potenziamento o superamento dell'umano?" *Acta Philosophica. Rivista internazionale di filosofia* 26 (II), 255–258. doi:10.19272/201700702001.

Rutigliano, E. (2001) *Teorie sociologiche classiche. Comte, Marx, Durkheim, Simmel, Weber, Pareto, Parsons.* Torino: Bollati Boringhieri.

Schleiermacher, F. E. D. (1799) "Versuch einer Theorie des geselligen Betragens." *Berlinisches Archiv der Zeit und ihres Geschmacksin* 5, 48–66.

Sessa, M. (2019) "Home smart home. A sociology of living in the age of reflective materialism." In Iannone, R., Gurashi, R., with Iannuzzi, I., de Ghantuz Cubbe, G. and Sessa, M., *Smart Society. A Sociological Perspective on Smart Living.* Abingdon: Routledge.

Shepard, M., ed., (2011) *Sentient City. Ubiquitous computing, architecture, and the future of urban space.* Cambridge MA: the MIT Press.

Shore, C. (1997) "Community." In Outhwaite, W., Bottomore, T., Gellner, E., Nisbet, R. and Touraine, A., eds., *Dizionario delle scienze sociali.* Milano: il Saggiatore, 107–108.

Spreafico, A. (2011) "Community." In Bettin Lattes, G. and Raffini, L., Manuale di sociologia, volume I. Padova: CEDAM.

Strassoldo, R. (1987) "Community." In Demarchi, F., Ellena, A. and Cattarinussi, B., eds., *Nuovo dizionario di sociologia.* Milano: Edizioni Paoline, 485–499.

Sztompka, P. (1996) *La fiducia nelle società post-comuniste. Una risorsa scomparsa.* Messina: Rubbettino.
Tönnies, F. (1979) *Comunità e società.* Milano: Edizioni di Comunità.
Volti, R. (2001) *Society and technological change.* New York: Worth.
Weber, M. (1922) *Wirtschaft und Gesellschaft.* Tübingen: Mohr.
Winthrop, R. H. (1991) "Community." In *Dictionary of concepts in cultural anthropology.* Westport: Greenwood Press, 40–43.
Zamagni, S. (2012) Bene comune e fraternità. *Il Contributo italiano alla storia del Pensiero – Economia.* Roma: Treccani.

5 Home smart home

A sociology of living in the age of reflective materialism

Melissa Sessa

Reflexive modernity: a critical analysis of the domotic drift

In these times of "worrying increase of urban population, scarcity of energy resources, high social conflict, ethical drift and disciplinary loss" (Fistola 2011: 74) the theme of the *smart home* with its corollaries polarizes the interest in multiple disciplinary and research fields. What does the house represent? Why did we feel the need to turn it into a conglomeration of technology? What are the effects on people and society?

The changes that have taken place with the new modernity have brought to the fore new needs that have acted as a matrix of the change of the house both to the material structure and to the relational one. A house that has turned from simple into *smart*, a house that has moved from tradition and materiality to the liveliness and interconnection with the technological world. Why did this change take place? Has smartness undermined the previous cultural bases to become the dominant paradigm?

The time frame taken as a reference, for the birth and development of the smart home, is reflexive modernity (Beck, Giddens and Lash 1999) which brought to the fore the environmental crisis with which we realized we could no longer understand the new challenges posed by the scarcity of resources in all its forms, through the traditional sociological categories. We found ourselves faced with the rising of the levels of individualization, with the crisis of alliances in social relations that are now being established "more by the conflict for the distribution of evils (pollution, danger) than for the division of goods" (Tacchi and Cucca 2003: 2). The house, in this climate, represents a social micro-cosmos in which all the problems that from the new modernity are reflected on the individual are represented. It is in the house that the individual shelters, it is on the house that the individual points his expectations, it is within the house that the individual builds his own life and then repeats it in the social macro-cosmos. The criticism of reflexive modernity is based precisely on how the old concept of modernity is understood as an opposite social form, by virtue of its characteristics, to traditional society (Mongili 2007), where technique and science on the one hand, and the company on the other are present. In the middle of this dichotomy we find the social actor and understanding the human being in his form of actor in space, in this modernity, means understanding how his needs and values have changed and how much these have affected his creation of home. In order to understand how this surrounding space has been shaped by man, we need to consider two closely related orders of ideas. The first is that man is not transversal to every epoch in history, and the second, consequently, is that it is historically determined, that is having a specific character and not given in kind once and for all, but every single time. In our case the one who will shape what is most structurally intimate, that is the house, will be the ideal type of smart people, which will create and modify the surrounding space with values and needs differentiated by degree.

The action of living in a house does not only mean being there at certain times of the day, but also means routinely traveling through its places, using its space and objects that come to life in that space, to satisfy one's needs. Specific needs that follow an ascending logic (Maslow 1987), from the simplest to the most complex, which characterize different motivations, and which differ according to the different types of objects to which they are linked. A staircase, a pyramid of needs that will reflect the needs, the experience of a home, which must not necessarily be considered a smart home, in its sense of a smart home, but rather a place that represents the result of the individual remaining in his indoor.

Security, declined in protection, belonging, dependence and stability is the most immediate of needs that recalls the need to guarantee both the elements that protect physical integrity and a stable and secure social dimension, whose solvency is preparatory to all others, or without which other needs find no way of manifesting themselves. As with all social phenomena, security can marry multiple definitions depending on the perspective from which it is observed, and in the sociological and psychological perspective it can be studied as a need connected to housing. Building a house, sheltering in something that is defined as a home by the inhabitant, that is solid, indoors, outdoors or that is not home for external subjects, is a necessity present in every man, in all the different historical periods.

It is no coincidence that security needs are defined as primary having a close link with biological needs and are met uniformly by people, regardless of the culture they belong to. The other needs are called secondary, characterized by a greater psychological connection and therefore by a higher degree of subjectivity. The first need of secondary order is marked by the appearance of the needs of affection, also declined as love, friendship, approval. Once the house is defined, the relationships within it will also be defined. The house is made up of places that are lived by the people who live there, who meet and collide. Intimate places where to consume relationships, as well as more open places where to experience collegiality. The need to forge relationships and to seek through them the acceptance by third parties, favoring an exchange, are the objectives of these needs. It is not by chance that the needs that follow those of affection come to be constituted by the different needs of esteem, relating to the need to build a positive self-image, which includes success, adequacy, mastery, respect, social position and appreciation. In terms of complexity and precisely for this reason the most difficult to reach are the needs of self-fulfillment, which correspond to the maximum development of individual abilities, and which shape the individual as it should be. This type of need can be traced in the construction of the ideal home, which does not necessarily mean the construction of a home that has always been desired, but that of a house that is suitable for one's life needs, aesthetically satisfying, which perfectly reflects the subject who lives there. Once one of the needs is satisfied, it will not generate more satisfaction until the gratification it generates will weaken, and it will reactivate it. The broad microsocial articulation of needs generated by Abraham Maslow will be reduced to a macrosocial dichotomy that separates materialistic needs on the one hand and post-materialist needs on the other (Inglehart 2015). The physiological and safety needs are brought within the first group, while the secondary needs can be traced within the second. With the expansion of mass communications, with increasing levels of education with the growing economic and technological development, it has been hypothesized as secondary objectives having been linked to objectives linked to primary needs. It is not by chance that the phenomenon of smartness linked to the political world is emerging on the social scene, with increasing arrogance, which allows a wide-ranging reasoning ranging from the economic-financial matrix (Geddes 2005) to the democratic matrix (Marciano 2018). This shift in values into needs has meant that the political and governance dimensions have also been linked to technology in a tight set of relationships that characterize the new society (de Ganthuz Cubbe 2019).

In fact, it is only thanks to horizontal subsidiarity and incentives on the part of the institutions that the creation of a smart home will be able to better contain the listed needs and will be able to better integrate into the smart society.

From home to smart home

The modern idea of home is "a scheme of articulation of the social space, so hidden, as sure of being shared" (Beck 2009: 42). An idea, therefore, of a global, and not globalized, home, where the description of enclosed spaces becomes fictitious because isolation is not possible. Analyzed in the context of space-time transformations, the idea of home is an integral part of what is called "modernist rationalization of space" (Bauman 2000), or of that overall control of social activities through delimitation, delineation and the division, in one word, across the border, either internal or external. At this point we can say that the house contains in itself a complex meaning and is therefore changeable over time, so that its borders cannot be given a priori, but are modified according to the relationship between the physical space and the meaning of from time to time society attributes to the relationships established there and which are determined by the prevailing culture (Mandich and Rampazi 2009: 10). There has always been a dual nature of the house: a physical, tangible and concrete reality defined as 'house', and an intangible reality, a structure of relationships, a way of being and feeling, defined 'home' (Coregliano 1991: 26). "House is therefore the basis for the development of a home experience" (Coregliano 1991: 23). The house belongs to the facts of life, it is a project of intimacy, a reflection on what will come and become from it. Home is a safe place, which protects from the anguish that is caused by the feeling of precariousness emanating from society. It is through the construction and the action in the house that the personality of the inhabitant with himself is revealed, and with what is external to the house. It is above all within the walls of the house that the intimate and affective life of people takes place (Mandich and Rampazi 2009: 2). We cannot define the house and understand where its boundaries are placed in the experience of the subjects, if we do not place ourselves in the interweaving of experiences, relationships and practices, which develops within it (Mandich and Rampazi 2009: 11). The lives of individuals are increasingly divided between an intimate sphere and a public sphere (Ariès and Duby 1988; Elias 1982), between secret behaviors and public action. In this direction, the division that is created between the sphere of personal life and the sphere of public life becomes a predominant feature of the new modernity and contains both the slow transformation of private relations and the confinement of emotions in this area of social life. Living in the house therefore becomes the expression of 'feeling at home', that is, recognizing oneself in the place where one is, in the objects that are handled, in the people one meets, in the activities one does. The house becomes a mental place, where the qualification of the house is expressed in the possibility of the subject to create his own centering in the world in which he lives. It is the result of what is defined by Agnes Heller as 'home world', and that rediscovers the meaning of the house as a dwelling (Heller 1999). Thus, the idea that the house is by definition "the theater of private life and more personal training" (Mandich and Rampazi 2009: 2) has become widespread in the imagination of Western societies, while what is and becomes external to it concerns the area of the public 'away from home'. A public sphere in which the subject finds a place in the measure in which he exercises a double role: the first as a member of a community, the second as a worker within that community (Mandich and Rampazi 2009: 4).

Simultaneously with the birth of the modern idea of domesticity, understood as the peculiarity of being within a space governed by logics completely separate from those imposed by the rationalization of public life, the concept of intimacy has developed, which, in common feeling, "has more and more linked to the ambit of affectivity, feelings, ablativity" (Mandich and Rampazi 2009: 4), in a single word, to the sphere of 'home'. The smart individual realizes that within the society of modern consumption, one interacts not only by dressing, but also by using furniture and the home itself. Building your own home, furnishing and living it, is an engaging experience because it means "making visible and communicating to others, but also to yourself, your own existential project" (Coregliano 1991: 113). It is a practical example of how building a house actually means living and communicating, we can find it within this book. The Smart Solar House ReStart4Smart is a green, smart, socially inclusive and integrated home with the surrounding environment, built and designed by young students, as will be seen, and which materially represents what has been said so far, which expresses as through simple planning and putting ideas into practice the subject communicates with the surrounding world (Gurashi, Iannuzzi and Sessa 2019).

So, from the simple home, we moved on to *smart home* technology. The advent of the new 4.0 society, of extreme contemporaneity, went on to develop a new housing phenomenology that laid its foundations on the *smart* concept. A magmatic concept that has always represented the desire to indicate an evolutionary state, attributable to the most recent developments in technological innovation (Fistola 2013). An unprecedented attention to *smartness* as a concept that represents something more than technological must be paid to sociology and its observation of facts and social actors. It is from the study of the organizational phenomenon that it has been possible to give a first precise definition of smartness (Fistola 2013) linking it to operations and materiality. However, it should be recorded that the existing literature on the smart home topic is very technical and tends to explain in detail how the presence of the grid (distribution calculation infrastructure) intersects the structure of the house, and not what this smart home is. Here we will try here to give a definition in the light of the concept of smartness. So, a home is smart (like any other device) when it is not only able to solve the problems that arise through the structuring of technological processes, but also when it is able to provide practical solutions for the solvency of those problems. It is no coincidence that the definition of smart home is almost like the definition of a smart community (Iannuzzi 2019), which presents the same etiological problems. If smart community focuses on the mediation between the needs of citizens, institutions and companies taking ICT (Information and Communication Technologies) as a basis for the quality of life of the inhabitants, in the same way a smart home makes its paradigm its own technology to improve the everyday life of its inhabitants.

And this is where the matrix of the change of the house is located, which undoubtedly can be traced to an intelligent use of technology. The house is configured in this way as a privileged observatory of the transformations that pass through everyday life and which increasingly see it intertwined with infrastructures and technological devices (Pellegrino 2011). Differently from electric houses, smart homes are able to help and facilitate the domestic routines of the subject who lives through a universe of sensors and actuators capable of monitoring and optimizing any activity and able to relate to the outside what happens to the 'indoor'. In this sense, with the development of modern culture the progressive confinement of the concept of 'home' within the physical walls of the dwelling, and the parallel polarization between places of privacy and places of public life that has established itself in the collective imagination, are nowadays challenged by a multiplicity of phenomena. The house in fact is structured as a progressively mediated space, which from an initial opening towards the outside, is articulated in an increasingly interconnected way to the techno and media branches (Appadurai 1996) that surround it and incorporate it into "pervasive communications networks" (Pellegrino 2011: 1). The mass media and the new information technologies have created large areas of permeability within the private space of the house that comes to be engraved and modified, and in the most extreme situations also inhabited by events and influences from the outside world, "altering the traditional physical-cultural criteria of closeness / distance, familiarity / strangeness" (Mandich and Rampazi 2009: 10). "The world is getting smaller and smaller and the bonds with what is far away are increasing, but at the same time it is getting bigger, since we are not able to dominate the new horizons that open up to our gaze" (Paino 2012: 2). Without a doubt, therefore, the transition from home to smart home is characterized by three elements. The first is represented, as we have seen, by the presence of technology, which has permeated the very essence of the house. The second element is found in the resolution of problems, which in the new smart home no longer represent an obstacle, but which are resolved by the house itself. And finally, the third element that makes the home a distributor of information from the inside out, making the boundary between the inside of the house and the outside, between the public and the private, ever thinner. But then, in light of what has been said, how do the structural characteristics of the new house change?

A look inside: the features of the smart home

We have seen that house and home are two different sides of the same coin, and how people taking over the house make it become home. We have also seen how there is not a univocal definition of the concept of smart home, but how much this, to be understood, must incorporate digital technologies and communication services (Gram-Hanssen and Darby 2016). Precisely for this reason, the STS theory (science, technology and society) that embraces the interrelations between the technical and social aspects of an organization is easily applicable to the *smart home* context. The focus is not only on the close connection between the term smart and the term intelligent, in the material meaning of it, but on the plots drawn from the connection that the term smart has with science and knowledge. It teaches us how technology, the progress of technology, if they are divorced from society, can generate dystonia. One cannot help noticing that what is created by technology, modified and at worst made obsolete, is inevitably "socially shape" (Williams and Edge 1996). "What could be said trivially is how much there is a need to understand the diffusion and use of technology, rather than its creation" (Gells 2004: 898). Technology has become a crucial element in modern societies, which are based on its innumerable functions, and of our house, which has become its main stage, which is why "it makes sense to distinguish between production, distribution and use made of technology with its functions and subfunctions" (Gells 2004: 900). In other words, it makes sense to make a precise analysis of the features that represent the smart home and give them a social matrix. These characteristics (comfort, solidity, flexibility, security, interconnection, sustainability, low cost and inclusion) in fact start from well-known social categories.

Comfort lends itself to the diverse needs of socializing of public private space. Which incorporates the new requests for accessibility that globalization brings with it, with the technological overcoming of fatigue caused by social impediments.

Solidity, understood as the durability of the house, is meant to be inserted in the context of generational sustainability. We tend to construct something that persists over time, not only structurally, but that preserves the surrounding environment for those who will live in that environment in a time t.1.

Flexibility, that is the ability to adapt the domotic envelope to the smart society, is here understood as social resilience, or rather as the ability to adapt despite the adversities, as the attitude of a community to establish a network of adaptive capacities as a result of a collective event that disturbs normality.

Security becomes confidence in the new generation systems, divided into safety and security. The plants conforming to the new industry 4.0 standards in the future will be able to guarantee safety, in the sense that they will have the possibility to reconfigure themselves and optimize themselves autonomously to guarantee a perfect yield.

Services offered by the smart home, also connected to the need to keep the cost of materials low, become inclusion, of the individual, and of the home, in the broader context of the smart society.

Internal and external *interconnection* is the engine of sociability. A smart home can be defined as fully functional only if all the innovations within it are able to interact and coexist in a systemic and dynamic way. By external interconnection, on the other hand, we mean the ability of a home automation system and its users inside, to interact with the outside, implementing the concepts of social life and community with the help of smart technologies, towards the primary purpose of quality of life. The home automation system is used to improve life inside the home, but it must not be limited to the home itself. We must look at a general dimension that also includes the other neighborhood houses, possibly smart ones too, in an interconnection perspective that allows the communication of all the systems. Thus the house shell is usually completed through one or more communication systems to the outside to allow the control and display of the status of the area monitored remotely, in order to possibly condition the sociality of the same smart homes between them.

Sustainability is understood both as environmental and social, as "smart home and smart grids will be part of an increasingly sustainable future" (Gram-Hanssen and Darby 2016: 2). In the current environmental emergency, the spread of a culture of sustainability takes the form of promotion, both at a personal and a social level, of styles of thinking and behavior based on savings such as optimal management of natural resources, but also as enhancement qualitative of human living, for a present, towards a near future. "The educational processes aimed at sustainability, promoted to guide the emerging participatory need, have as their main purpose the increase of the capacity of the community to be resilient" (Beretta 2015: 91).

The *low cost* traces the footprint of social availability. An essential feature of the home automation system must be that of low cost, understood as the installation of technologies that have affordable prices for all. It is about including the generality of users by giving them a wide choice of smart solutions. Today, the cost of a building must also be measured by taking into consideration its entire life cycle. In addition to the incidence of the purchase value of the home, it is necessary to consider the management costs of the same, which sometimes exceed the cost. So, the concept of smart home can be summarized in four large macro categories: such as security and control; place of actions and practices; indicator of a social status (Gram-Hanssen and Darby 2016); and finally, as inclusion and interconnection. As for the first category, this identifies the house as the center of control. Inside the housing envelope the subject feels protected, not only thanks to the personalization of this according to his own needs, but also from the external society, sometimes perceived as hostile (Gram-Hanssen and Darby 2016). The concept of safety, defined as safety and security, plays a prominent role. The smart home is identified as "conflict free" (Gram-Hanssen and Darby 2016: 5). New technologies make it possible to reduce risks to a minimum, are designed and structured in order to have minimal impact on the environment, and to optimize the activities of the actor. The possibility of continuously monitoring, recording and tracking all the events that occur within a domestic environment has the advantage of favoring an objectivity of evaluation that can be considered as a rating, the complete exclusion of a value judgment. In many cases, behavioral changes are so slow and gradual that they are imperceptible to the human eye, so there is a need for technologies that help capture these differences. It is also an economic question of trade-offs where factors such as cost, invasiveness and privacy come into play and must be weighed in the most appropriate manner according to the need and the objective to be pursued. What was once a simple manual switch, is replaced inside the smart home with an electronic switch that acts as a sensor or a local multi-function actuator, which not only regulates the quantity of the emission, but also the quality of the same.

The house is also a place of actions that are recorded in everyday life, from cooking, to eating, to sleeping. It is no coincidence that the social actor shapes the *home* in such a way that it can be defined as the center of everyday life. The same domotic systems already described as multi-function actuators have the possibility to learn independently the everyday life of the inhabitant.

The third point attributes to the house a role of indicator of social status. The house is a property and as such reflects the ideas and values of those who live there, temporarily or permanently. The decorations of the house not only show others who the subject that inhabits the place is but are also the reflection of who the subject is for himself, of how 'aesthetic' self-realization can pass through the aesthetics of the home. What people do with their homes, in the form of supplies or decorations, allows us to understand the different cultures of consumers.

The fourth and final point is the definition of the smart home as inclusion and interconnection. A smart home can be defined as fully functional only if all the innovations within it are able to interact and coexist in a systemic and dynamic way, to the extent that the term smart is emphasized in its sense of human intelligence. This intelligence must be stimulated within a process that aims at inclusion, undoubtedly a key element of this new way of living at home. At the heart of the smart home challenge is, yes, the construction of a new kind of common good, a large technological and immaterial infrastructure, albeit in materiality, which allows people and objects to communicate, integrating information. In the smart home context, there is above all the generation of intelligence and the production of inclusion in order to improve the domestic routine, to optimize everyday life. The objective therefore is to make the environment in which we live interactive and cooperative. Effective and efficient in supporting the independent life, able to provide greater security, simplicity and satisfaction in carrying out the activities of daily life. "Inclusion means avoiding all forms of discrimination and valuing differences" (Beretta 2015: 62).

Reflection points: the ethical responsibility of smart home

After reviewing the characteristics that make the smart home an agglomeration, an intelligent and social technology, we'll move on to analyze the problem of time. The houses are built to last, but as these homes are technological, and with technology being subject to obsolescence, I think we should dwell in this section on the importance of the time factor, and on how this transforms our smart home. Certainly, as said, "if we study the problem of time, we can learn things about men and, therefore, also about ourselves that we previously ignored. Problems become accessible to sociology and the human sciences in general, given the stage so far reached by their theory, still remain incomprehensible" (Elias 1986: 7). The house has traditionally been studied as a center of spatial orientation, that is where man was, and of the social order, or that man's place occupied. More rarely, on the other hand, it has been related to the integrative dimension of identities and to the temporal orientation, understood as the subject's ability to construct his own personal history and place himself in it.

What does building mean? And what do you live in? To be able to understand the relationship between these two actions it is necessary to understand that the action of living is precisely an action performed by the social actor, which identifies the 'I live' with the 'I am'. Only if you have the capacity to live, will you be able to build a house both materially and subjectively. Making a place a place is a process that requires planning and investment (Pozzi 2015) and only if you have the capacity (not least materials) to invest and plan will you also be able to place yourself in a place to define it home. By focusing on the relationship between *home*, past, present and future, or the three components of biographical temporality, one can see how the house is traditionally configured as a basic element for the affirmation of "permanence, continuity and coherence biographer of its inhabitants" (Coregliano 1991: 14). Starting from the house as "memory of past experience, way of being in the present and projection towards the future" (Coregliano 1991: 14), it will be possible to arrive at the study of how these realities modify the dimensions constituting a new binomial, that between individual and living space. How does the individual fit into the space of the house? The following reasoning will try to trace a red thread that invests and makes three factors that apparently might not seem to be connected: the smart home, temporality and efficiency.

It is possible to notice how the dimension of the present is perceived as flattened and restricted by considering that today the temporal perspective loses some of its characteristics based on coherence and continuity. The speed with which information, people and objects move, inevitably limits the perception that the social actor has of his time, and of the time that surrounds him. The action of the present untying itself from a precise project, which would imply the future, allows us to experiment ways of acting unencumbered, acquiring new values, and giving the possibility to the subject to concentrate most of his energies on it. "The concentration of modernity on the perspective of the present has hidden the perspective towards the future" (Nocenzi 2011: 1). "The future cannot be" read the title of one of the essays by Niklas Luhmann (1976), which emphasized how modernity tended to simplify, to streamline the difficulties of present reality, depriving it of future implications, just starting from the effects of the actions taking place in the present. On the other hand, the social sciences speak of the future by overcoming the temporal scan between the present and its past. By changing the connection between what happened in the past, what happens in the present, and what will happen in the future, the organization of the everyday inevitably changes, free from the chains of the past, and from the idea of the future, acquires new meanings linked to contingency and contemporaneity. A newspaper closely linked to the concepts of effectiveness and efficiency that allow it to be a cutting-edge present, because when speaking of smart home, we have seen how this is closely linked to three factors: technology, interconnection and problem solving. Precisely for this reason, aimed at fully analyzing the essence of the smart home over time it is necessary not to forget how efficient and effective this should be. If we intend to make a temporal analysis of the smart home, it is also necessary to point out how this mantra makes perfection. A house built to last, designed to give a future perspective, which is not effective and efficient cannot be conceived. Consequently, the concept of efficiency is of primary importance when it comes to the smart home, since the home automation project must be able to increasingly guarantee any service that is equipped. Efficiency must be distinguished from effectiveness, while the former is based on a cost-benefit ratio, the latter looks at the set objectives. People will always find themselves in front of something efficient, which is not equally effective. Thus, the smart home marries both principles when the objectives set, whether they are construction standards or rather capabilities that the inhabitant should acquire, go hand in hand with leveling the costs of the product on benefits for the person, and in a vision more general than society. It is a new form of consumption with more blurred boundaries, which becomes communication and language. One cannot help but notice how, even through the home, the concept of capitalism and the capital invested have changed. Importance is given to immaterial capital, qualified today as human capital, knowledge capital or intelligence capital. "Knowledge is made up of experiences and practices that have become intuitive evidences and habits, and intelligence covers the whole range of capacities ranging from judgment and discernment to openness, to the aptitude to assimilate new knowledge and combine with knowledge" (Gorz 2003: 11).

A different economic rationality stands out in front of the society that is no longer based on the criteria of performance subordinated to human development. The overcoming of this type of capitalism, marked by a purely material connotation, must pass behind the overcoming of productivism. We reason in terms of a hypothesis in which the new economy would undress the clothes of its domination over human capacities and qualities cease to be a means to wealth but were wealth itself. Production should be put at the service of man, and not the other way around, underlining in an even more marked way the clear difference that exists between producing and producing, where the acting subject emancipates itself to become capital itself. That is, one should note the sad connection, already pointed out and described by Marx, between the world of men and the world of objects, where "the devaluation of the human world grows in direct relationship with the enhancement of the world of things"

(Marx 1844: 71). The technological fetishism that accompanies the growth of the smart home is obscure as the basis of technology is man, bringing to light the need to build houses that are not always more technological simply for the sake of being, but that are as much technological as well as human. By this I mean that if the use of ICTs in the home is not linked to the needs of the individual rather than to the needs of the market, we will often find ourselves in front of houses that in the abstract could satisfy almost all users, but that on test with reality the connection with the individual fails.

Conclusion: is it time for new values?

At the end of this sociological-descriptive path on the smart home, defined in its entity and its characteristics, numerous questions have emerged. Has the advent of such a modern, intelligent home changed the values of the people who live there? In a world where everything is foreign to everything, has the need for security come to be increased? Is building a solid home really the answer to the attacks on society?

We have seen how the home is not a place alienated from the society in which it has fallen but is an integral part of a context that increasingly pushes towards a synergy between the technological components and the social components. How the word *home*, with its strong intimate and emotional meaning, can tame the word *smart* and the two terms can flow together in a combination of intelligence and intimacy to create a satisfying hybrid that represents the inhabitant. We then noticed how the culture of living shapes the subject's vision of the world so much that one could say that every person makes a place. In light of the above, not only is the world perceived starting from the phenomenological experience of one's living space, but this space is given meaning and structured according to one's own perspective. A living perspective that expresses its characteristics in the connection with the cultural organization of society that is reflected within and outside the home. Therefore, a radical change in the perception of the house and its values becomes inevitable for people, who no longer recognize the smart home as an exception to normality. The essence of the term normally used is re-established, meaning that the technological house is more present than expected, and how, perhaps, the reader at the beginning of this essay could have thought. Just as during the Second World War the house was no longer defined as an electric one, but simply as a home, due to the normality of electrical systems inside homes, so "the house that we now call intelligent in a few decades will be the average standard of installation " (Capolla 2011: 57), that is, the production of home automation systems begins to take on a configuration suited to a possible and non-futuristic domotics. To something that really is and could not be. In the routine of creating the smart home, innovative standards are beginning to be set up, and are no longer new, which make the perception of the home shift as little as possible, but as actually realized. Values such as those of domesticity, multi-functionality and housing flexibility are rediscovered. The same domestic everyday life characterized by growing uncertainty in the temporal, spatial and relational definition analyzed in the previous paragraph, reflects "the complex intertwining between individuality and collectivity, between personal intimacy and family intimacy" (Mandich and Rampazi 2009: 14) that take life within the shared space of the home. Secondly, the increasing multi-functionality that is, with the advent of new technologies, increasingly characterizing the space of the house-dwelling, has made increasing levels of flexibility that have developed within the inhabited space possible. A flexibility that by its nature is a source of negotiations and sometimes even of continuous tensions between those who frequent the same spaces of the house. Thus, in the modern imagination, the smart home is characterized by its technological resilience, which allows it to adapt to the needs of the subjects passing through its spaces, and which, as we have seen, can be a meeting place and a place of tension.

Despite the scientific dictates showing the technological power as characterization of the smart home, it should be noted that the idea of home, in the traced descriptions, supports also and above all the first need analyzed: safety. Hence the idea of a home that is *safety*, that is, that foresees that the residual risks related to a machine or plant do not exceed acceptable values, and of a house that is *security* or that protects from unauthorized access from outside, of whatever nature they are. With the advent of the new 4.0 society we noticed how the fluidity of accesses, of knowledge, the interconnection between technologies and social actors, made security prevail over safety, over the guarantee of good functionality. There is more concern than guaranteeing standard results, protecting ourselves from attacks that may arise from external users. It is no coincidence that in recent years hacker attacks, from the infringement of Pentagon servers, up to the most trivial violations, have been increasing. We try, in every way, to enter the life of the stranger, to sneak into the different from himself, to bring to light the need for a synergy between the matrix technology and the social matrix. A smart home can be fully inclusive and integrated only if it is in a social context that is smart itself and that allows it to be secure. In fact, we can see how the growing transformation of the society into *smart society* has allowed the spread of the smart housing model more easily. It is within the new society, through integration, inclusion, governance and people that any smart initiative can be made possible (Iannone 2019).

It is difficult to find even a single place, where objects exist without being 'full of people', in the same way no society that claims to be human can function insubstantially, without relying on both material and technology, and likewise no society can be called smart if it does not contain, within it, as many types of technologies that are smart. Those same technologies that allow the mediation between men and things and that include within them a set of heterogeneous actors. Those same technologies that are driving a change that brings the eye to the future and to the new generations that will have to deal with the new smart needs. It is never possible to consider the house, as well as the domestic spaces that are contained within it as a place in itself, in isolation, which can be detached from urban spaces and larger social spaces. Instead, it is necessary to look at the forms of social integration of this house, in the spatial organization that encompasses it, and within society.

The originality of the smart home is therefore to be sought more than in technologies, in the people who live there, in the smart people (Gurashi 2019) who are not only consumers but also producers of their consumption, those who are called 'prosumers' (Degli Esposti 2015). Undoubtedly, we must consider the attractiveness of the 'new' as the main push towards the perception of the subject not only as a spectator, but also as an actor. A subject that has evolved from a simple consumer to a producer, up to even questioning this dichotomy to become a 'prosumer'. And it is here that we have noticed how the tendency is established to seek a relationship between the social structure that inhabits the smart home and the city that hosts it, since the *smartness* of a city is given by people more than by the technologies that make it up.

In conclusion, with a development of the house that defines itself as smart, we do not only think of the present generations, of safeguarding the present, but also and above all of protecting a future, not even so near. The smart home represents a small step towards a future that is certainly easier, if we can perceive the house as a bigger system and that smartness is a daily life for improving the quality of life.

References

Appadurai, A. (1996) *Modernità in polvere*. Milano: Raffaello Cortina Editore.
Ariès, P. and Duby, G. (1988) *a cura di, La vita privata. L'ottocento e il Novecento*. Bari: Laterza.
Bauman, Z. (2000) *Modernità liquida*. Bari: Laterza.

Beck, U. (2009) *Che cos'è la globalizzazione. Rischi e prospettive della società planetaria*. Roma: Carocci Editore.

Beck, U., Giddens, A. and Lash, T. (1999) *Modernizzazione Riflessiva. Politica, tradizione ed estetica nell'ordine sociale della modernità*. Trieste: Asterios.

Beretta, I. (2015) (eds.) *L'umanesimo della smart city, Inclusione, innovazione, formazione*. Lecce: Pensa MultiMedia Editore.

Capolla, M. (2011) *Progettare la domotica. Criteri e tecniche per la progettazione della casa intelligente*. Santarcangelo di Romagna: Maggioli Editore.

Coregliano, M. (1991) *No place like home*. Milano: FrancoAngeli.

Degli Esposti, P. (2015). *Essere prosumer nella società digitale: produzione e consumo tra atomi e bit*. Milano: FrancoAngeli.

de Ghantuz Cubbe, G. (2019) "Smart Politics. The Political Dimension of the 'smartness'." In Iannone, R., Gurashi, R., with Iannuzzi, I., de Ghantuz Cubbe, G. and Sessa, M., *Smart Society. A Sociological Perspective on Smart Living*. Abingdon: Routledge.

Elias, N. (1982) *La civiltà delle buone maniere*. Bologna: Il Mulino.

Elias, N. (1986) *Saggio sul tempo*. Bologna: Il Mulino.

Fistola, R. (2011) "Urbanistica e Pianificazione fra crisi ed Innovazione." *Urbanistica Informazioni* 235.

Fistola, R. (2013) "Smart city. Riflessioni sull'intelligenza urbana." *TeMa Journal of Land Use, Mobility and Environment* 6(1) 47–71. doi:10.6092/1970–9870/1460.

Geddes, M. (2005) "Neoliberalism and Local Governance – Cross-National Perspectives and Speculations." *Policy Studies* 26(34), 359–377. doi:10.1080/01442870500198429.

Gells, W. F. (2004) "From Sectoral Systems of Innovation to Socio-Technical Systems. Insight about Dynamics and Change from Sociology and Institutional Theory." *Research Policy* 33, 897–920.

Gorz, A. (2003) *L'immateriale, conoscenza valore e capitale*. Torino: Bollati Boringhieri.

Gram-Hanssen, K. and Darby, S. J. (2016) "Are 'Home' and 'Smart' Contradictory Concept or Fluid Position that Will Converge?" Paper prepared for DEMAND Centre Conference, Lancaster, April 13–15, 2016.

Gurashi, R., Iannuzzi, I. and Sessa, M. (2019) "From theory to empiricism: the challenge to the future of the Sapienza University at Solar Decathlon Middle East." In Iannone, R., Gurashi, R., with Iannuzzi, I., de Ghantuz Cubbe, G. and Sessa, M., *Smart Society. A Sociological Perspective on Smart Living*. Abingdon: Routledge.

Gurashi, R. (2019) "Smart people: the individual challenge of the fourth industrial revolution." In Iannone, R., Gurashi, R., with Iannuzzi, I., de Ghantuz Cubbe, G. and Sessa, M., *Smart Society. A Sociological Perspective on Smart Living*. Abingdon: Routledge.

Heller, A. (1999) *Dove siamo a casa? Pisan Lectures 1993–1998*. Milano: FrancoAngeli.

Iannone, R. (2019) "Smart Society. The critical sense of a world strategy." In Iannone, R., Gurashi, R., with Iannuzzi, I., de Ghantuz Cubbe, G. and Sessa, M., *Smart Society. A Sociological Perspective on Smart Living*. Abingdon: Routledge.

Iannuzzi, I. (2019) "Smart community: a new way of being together?" In Iannone, R., Gurashi, R., with Iannuzzi, I., de Ghantuz Cubbe, G. and Sessa, M., *Smart Society. A Sociological Perspective on Smart Living*. Abingdon: Routledge.

Inglehart, R. (2015) *The Silent Revolution. Changing Values and Political Styles Among Western Publics*. Priceton: Priceton University Press.

Latour, B. (2015) *Non siamo mai stati moderni*. Milano: Eleuthèra.

Luhmann, N. (1976) "The Future Cannot Being: The Temporal Structure in Modern Society." *Social Research* 43(1) 130–153.

Mandich, G. and Rampazi, M. (2009) "Domesticità e addomesticamento. La costruzione della sfera domestica nella vita quotidiana." *Sociologia@Dres Quaderni di Ricerca* 9(1) 1–30.

Marciano, C. (2018) "Democrazia locale e media digitali. Verso una critica della Smart City." In Vitale, E. and Cattaneo, F. (eds.) *Web e società democratica: un matrimonio difficile*. Torino: Accademia University Press.

Marx, K. (1968) *Manoscritti economico-filosofici del 1844*. Torino: Einaudi.

Maslow, A. (1987) *Motivation and Personality (Third Ed.)*. New York: Addison-Wesley.

Mongili, A. (2007) *Tecnologia e società*. Roma: Carocci Editore.

Nocenzi, M. (2011) "Sostenibilità e dimensione generazionale." *Rivista di studi sulla sostenibilità* 1, 75–91. doi:10.3280/RISS2011–00101.

Paino, R. (2012) "La globalizzazione, lessico e significati. Brevi note sul dibattito nelle scienze sociali." *Quaderni di Intercultura* anno IV, 1–11. doi:10.3271/N40.

Pellegrino, G. (2011) "Nuove domesticità: la casa connessa e le tecnologie pervasive per la mobilità." *M@gm@* 9(3).

Pozzi, G. (2015) "Heidegger ai margini. Antropologia e trasgression." *Phylosophy Kitchen*, 2(2) 95–109.

Rampazi, M. (2007) "I mutevoli confini della domesticità nello spazio-tempo contemporaneo." Seminar on "The daily construction of domesticity. Transformations and continuity," Padova.

Rifkin, J. (2014) *La società a costo marginale zero. L'internet delle cose, l'ascesa del commons collaborativo e l'eclissi del capitalismo*. Milano: Mondadori.

Tacchi, E. M. and Cucca, R. (2003) "Modernizzazione riflessiva, ambiente, partecipazione sociale." Proceedings of the IV National Conference of Environmental Sociologists, Torino.

Williams, R. and Edge, D. (1996) "The Social Shaping of Technology." *Research Policy* 25(6) 865–899. doi:10.1016/0048–7333(96)00885–00882.

6 From theory to empiricism

The challenge to the future of the Sapienza University at Solar Decathlon Middle East

Romina Gurashi, Ilaria Iannuzzi and Melissa Sessa[1]

Introduction

The aim of this chapter is to analyze, through an empirical case, how the different dimensions of *smartness* come together and integrate with each other, considering the broader context provided by the European Union through the Horizon 2020 program. Reference will be made to the experience of the Sapienza team competing in the international Solar Decathlon Middle East 2018 competition, which will be discussed later.

The European Union, with the Horizon 2020 project, aims at favoring safe, innovative, and inclusive societies that progressively reduce energy consumption. In this context, the international Solar Decathlon competition, adopting the values of sustainability and energy efficiency as primary objectives to be achieved, has launched a challenge among the universities of the world for the construction of an energy self-sufficient home. The Sapienza University has taken up this challenge by presenting its ReStart4Smart project, a revolutionary architectural model 4.0 of Smart Solar House, equipped with the most advanced technologies and able to meet the requirements of efficiency, safety, comfort, convenience and sustainability dear to the 20th-century prosumers.

In addition to being an example of a *smart home*, the ReStart4Smart project is developed by incorporating the other dimensions of smartness into it. In this sense, a central role is played by the inhabitant of the house as an individual, as a bearer of needs, as a consumer and as an ideal type of *smart people*; in the guise of being-in-relationship – as a subject that relates constantly with other subjects, calling into question the community dimension (*smart community*); and, finally, as a citizen, or a subject that constantly relates to the political sphere and institutions, which today is becoming a smart *politics*. Everything, of course, is designed to ensure that smartness informs itself of the broader social level: the whole society (*smart society*). The Restart4Smart case therefore represents a privileged stage for analyzing these dimensions in mutual interaction.

The evolution of society towards an intelligent perspective has, as is known, reoriented the needs of individuals, politics, the community, and the home. The smart home ReStart4Smart is designed and built to be inserted within a social context capable of amplifying its functions, in a framework that tends to establish a relationship of dialogue with society and its social structure. This case study allows, therefore, to critically analyze the relationships between all the aspects previously studied. It therefore constitutes a real laboratory in which to experiment practices related to the facets of the 'intelligent' phenomenon, not only in their individuality, but also in their mutual correlations: in particular, considering the practices related to the actions of intelligent people in the context of a home intelligent, in relation to the size of the intelligent community and intelligent politics, within a society that aims to be intelligent as a whole.

A European vision: Horizon 2020

A fundamental tool of analysis and debate on the main research directions underlying the design of the Smart Solar House consisted in the study of the European vision on sustainable development present in Horizon 2020. The European Union, in fact, has always shown itself as attentive to environmental issues and energy, since its foundation. One of the most important projects set up in favor of sustainable energy fuels and in favor of energy efficiency and rationalization is Horizon 2020. As is known, it is the European Commission's instrument for financing scientific research and innovation, the European Union's executive body, which has allocated a budget of about 80 billion euros for about seven years, from 2014 to 2020, bringing together research funding[2] in some sectors, with an increase of about 23% compared to the previous phase.[3] This program today funds research projects or actions aimed at scientific and technological innovations that have a significant impact on the lives of citizens. It is aimed at a calibrated use of efficiencies. Energy efficiency is seen as a generator of added value. Innovation as true progress when it is within everyone's reach. It pursues the objectives of the Europe 2020 strategy and of the 'Innovation Union' initiative, that is smart, inclusive, and sustainable to increase Europe's global competitiveness. Within the Horizon 2020 program it is therefore possible to see the dimensions that have been identified in this work as different facets of the so-called *smartness*. Horizon 2020 overall refers to a series of fundamental aspects for the affirmation and development of a *smart* company. A company that is smart because it is composed of spheres and *smart* subjects in their turn. It sees, that is, within it the presence of a *smart* managed policy, which is aimed at *smart* communities composed of *smart* people who live in *smart* homes.

The present survey, in particular, highlights how what was hoped for by Horizon 2020 acquires meaning in terms not only of *smartness* in general, but of smartness declined according to the perspective of social integration. This emerges clearly where the growth to be pursued is identified, among others, with the adjective 'inclusive'.

The desired *smartness* therefore takes on multiple connotations, particularly in the three pillars – articulated internally in specific programs and themes – on which Horizon 2020 is founded (http://www.ap re.it/ricerca-europea/horizon-2020/).

'*Scientific excellence*' is the first of the three pillars that focuses on the level of excellence of the European scientific base to ensure long-term global competitiveness. Specific actions are foreseen to support researchers who want to implement their projects in Europe. Included within this pillar are: frontier research funded by the European Research Council which is a funding for researchers who propose innovative and creative research projects; Maria Sklodowska-Curie actions, a mobility program for young researchers, in order to favor international training opportunities; future and emerging technologies to support large and highly innovative projects for the creation of new technologies; and finally world-class infrastructure with actions aimed at supporting the latest and most advanced technological infrastructures.

The second pillar is that of '*industrial leadership*' that wants to bring Europe's investments into industrial technologies by guaranteeing adequate funding. Actions are presented to strengthen the capacity for industrial and business development of companies, innovations with high potential for technological development, and information and communication technologies. Within this pillar there are three strands of intervention: leadership in support and industrial technologies with technological support for innovation in all sectors, especially for ICT (information and communication technologies) technologies for space, nanotechnologies, production of advanced materials, biotechnology and highly specialized manufacturing industry; access to risk finance that promotes access to finance for new high-risk ideas and their development, and support for actions to attract private funding and venture capital for research and innovation; innovation in small and medium enterprises with ad hoc actions aimed at supporting the development and internationalization of SMEs.

And finally, the third pillar that concerns the '*challenges for society*' where social sciences and the humanities are an integral part of the activities aimed at addressing the strategic priorities of the Union and reflect the concerns of citizens. With the actions within this pillar, an approach is taken to integrate resources from all fields of research, including social sciences. Within this pillar the Union has identified seven priority challenges that are broken down as follows: health, demographic change and well-being; food security, sustainable agriculture and forestry, marine and maritime research and inland waters, and bio economy; safe, clean and efficient energy; intelligent, green and integrated transport; climate action, environment, resource efficiency and raw materials; Europe in a changing world – inclusive, innovative and reflective societies; safe societies – protecting the freedom and security of Europe and its citizens.

The spread of excellence and the promotion of participation, science with and for the Company, European Institute of Innovation and Technology (EIT), Euratom and Actions of the Joint Research Center are other cross – cutting activities of Horizon 2020.

The aim is to create safe, innovative, and inclusive societies that progressively reduce energy consumption. The goal is to move from a concept of expenditure to an investment concept, opening up to new types of sustainable energy markets, creating user-friendly technologies in order to support and increase the birth of smart cities. The focus on smart cities will translate into commercial-scale solutions with a high-potential market (http://ec.europa.eu/programmes/horizon2020/en/h2020-section/secure-clean-and-efficient-energy).

The new challenges that today's society is called upon to face, and which concern its ability to become ever *smarter*, fit fully within the third pillar of the Horizon 2020 program. The *smartness* of the company, as mentioned previously, comes to be defined, in this sense, in terms of the objective of a smart, sustainable, and inclusive growth.

This objective is strongly linked to the use of technologies able to stimulate responsible behavior and to the idea, carried out within Europe, of the integration between the different social sectors in terms of policies to be implemented. From this point of view, the reflection on sustainability cannot be separated from the role of scientific excellence and industrial leadership. The construction of the 'knowledge and innovation community' to establish new approaches to sustainable growth connected to entrepreneurship emerges as central. *According to Horizon 2020, sustainable development must not, in fact, be understood solely as responsible management of natural resources, but also as a social and economic responsibility, meaning the concept of sustainable development in the broadest sense of the term.*

The sixth challenge of the third pillar, 'Europe in a changing world – inclusive, innovative and reflective societies', intends to take into account, in particular, these needs, aimed at promoting solutions and supporting inclusive, innovative and reflective societies, in a context of unprecedented transformations and growing global interdependencies (https://ec.europa.eu/programmes/horizon2020/en/area/social-science s-humanities).

The degree of innovativeness of a society therefore appears to be closely connected to the level of inclusiveness and reflectivity of the company itself. This way of understanding the social reflects the fuller sense of *smartness* that a society can manifest, there can be no innovation – or technological progress – if it is not accompanied by a high level of inclusiveness, if, therefore, the objectives of the *smart society* are not pursued in a systemic and integrated manner.

From this point of view, innovation – strengthening of the knowledge base and measures to support the Innovation Union and the European Research Area (ERA), the search for new forms of innovation, including innovation and social creativity, guarantees of the participation of society in research and innovation, the promotion of a coherent and effective collaboration with third countries – travels hand in hand with the inclusiveness – construction of flexible societies in Europe, an increase in the role played by Europe on the world stage; a solution to the research and innovation gaps in Europe.

The role of reflexivity that society can exercise is also fundamental, as mentioned above. It is also through the analysis of memory and identity that it is possible to achieve a high level of integration and, therefore, guarantee one of the most important goals of the *smart society*, inclusion. Through this pillar of Horizon 2020, in fact, it emerges that the *smartness* of a society cannot be declined exclusively according to a perspective that is attentive to technologies or to technical-engineering aspects only. Just think of how important agencies, such as, for example, Enea (the National Agency for Technologies, Energy and Sustainable Economic Development), provide fundamental technical engineering details, but do not sufficiently consider the social problems arising from interfacing with new technologies and problems of sustainable energy consumption.

At the center of attention are thus placed, at least in terms of intentions and basic value orientations, all those economic and non-economic mechanisms (of a psychological, social and cultural nature) that are at the base of the choices of individuals and that bring the balance needle hanging in favor of new sustainable lifestyles.

The competition: Solar Decathlon Middle East 2018

The Solar Decathlon is a biennial international competition of architecture, engineering, and design, conceived by the United States Department of Energy (DOE) in 2002 and featuring universities from all over the world as protagonists. Among the objectives of the competition stands the development of innovation and the dissemination of a broad knowledge of renewable energy and sustainable architecture (https://www.solardecathlon.gov).

The competition, in fact, consists in the design and construction of a prototype of a house of about 100 square meters, completely fed, in terms of energy, through renewable sources. The entire competition covers the time span of two years, during which the various teams are called upon to face numerous challenges and difficulties. For this reason, the teams are always multidisciplinary, that is, they are made up of students from the faculties of architecture, engineering, and social sciences and communication. At the end of the two years, each team transfers their home prototype to a chosen location at the beginning of the competition and there, for an entire month, it will be tested in its ability to satisfy different needs, including its degree of effectiveness, of efficiency, environmental quality, functionality, comfort and so on (https://www.solardecathlon.gov/).

There are ten contests – hence the term 'decathlon' – which each team must address and by which the winning team is decided: architecture, engineering and construction, energy management, energy efficiency, comfort, functionality, sustainable mobility, sustainability, communication and innovation. Very important, in fact, is the role of communication, which therefore reflects the team's ability to know how to disseminate knowledge related to environmental issues related to the competition. In this sense, it is fundamental to consider the increase in awareness not only of the participants in the competition, but also of the sector of construction professionals, to ensure that they select eco-friendly materials, considerably reducing the environmental impact of housing and highlighting how a high level of energy efficiency is not necessarily in contrast with the needs of comfort and architectural beauty.

Since the Solar Decathlon is a competition reserved only to universities, it represents a real gym for students, who can put into practice the knowledge acquired in a continuous and permanent laboratory for a period of two years, comparing themselves within a stimulating and encouraging context such as the international one. It is for this reason that the Solar Decathlon is much more than a simple competition. It is an intensive learning experience that involves the aspects connected to the consumption and management of the home and that allows interfacing with the most recent and advanced home automation technologies and solutions for the home, including, for example, conservation measures and water recycling.

Starting in 2002, the Solar Decathlon has involved more than 150 multidisciplinary teams, has established itself worldwide as a successful educational program and as an opportunity to develop the working capacity of thousands of students.

The uniqueness of this competition has decreed its success over the years, leading to the conclusion of agreements between the US Department of Energy and numerous territories, aimed at being able to host subsequent editions of the competition.

Many have taken place all over the world, including: Solar Decathlon Europe in 2010 and 2012 in Madrid, Spain; Solar Decathlon Europe 2014 in Versailles, France; Solar Decathlon China 2013, in Datong; Solar Decathlon Latin America and the Caribbean 2015, in Santiago de Cali, Colombia, until reaching the 2016 edition called *Solar Decathlon Middle East*, ended in Dubai, United Arab Emirates, in November 2018 (https://www.solardecathlon.gov/about.html).

The announcement of the *Solar Decathlon Middle East* was officially released during the opening ceremony of the World Green Economy Summit which takes place every year in Dubai and took place in the presence, among others, of the UAE Minister for Climate Change, Thani Al Zeyoudi, of the Director of the United Nations Development Program, Helen Clark, of the General Director of the International Renewable Energy Agency, Adnan Z. Amin, and of the Consul General of Italy in Dubai, Valentina Setta. The competition started with the agreement concluded between the Dubai Water and Electricity Authority (DEWA) and the United States Department of Energy (http://www.architettura. uniroma1.it/archivionotizie/solar-decathlon-middle-east-#2018/4).

This is the first edition of the Solar Decathlon which took place in the Middle East and which, therefore, saw the teams face the difficulties arising from the critical local climatic conditions, characterized by extremely high temperature and humidity values. Twenty-two university teams[4], from 16 different countries selected by an international jury from over 60 candidates, through a rigorous technical review process lasting over three months took part in the competition. In addition to the 10 planned contests, the Solar Decathlon Middle East has been provided with four other special prizes: interior design, photovoltaic integration, technological innovation, public acceptance.

The *Solar Decathlon Middle East 2018* was won by the Virginia Tech University of the United States, which overall obtained the highest score, deriving from the sum of the scores for each contest, excelling, in particular, in the fields of architecture, energy efficiency with regard to sustainable transport, sustainability and, finally, in the field of innovation in engineering and construction.

As for the future, in 2019 three Solar Decathlon competitions will take place: Solar Decathlon Europe, in Szentendre, Hungary, in July; Solar Decathlon Latin America and the Caribbean in Colombia, in December and, finally, Solar Decathlon Morocco in September (http s://www.solardecathlon.gov/about.html). Therefore, there are numerous challenges, but also the opportunities that the Solar Decathlon will continue to place on university students all over the world.

The case study: ReStart4Smart

ReStart4Smart[5] is a project of the Sapienza University of Rome that has managed to be part of the small international group of 22 universities, coming from 16 countries, competing at the Solar Decathlon Middle East 2018, at its first absolute participation in this competition.

Immediately after the design and material construction, which took two years, the Solar House was disassembled and reassembled to carry out tests for procedures and construction time simulations in view of the final stage of the competition that took place in Dubai from November 14th to 28th at the Solar Hai, inside the Mohammed bin Rashid Al Maktoum Solar Park, the largest photovoltaic park in the world. The project was supported by MIUR, the Ministry of Education, University and Scientific Research, but also by more than 45 leading companies in the sector, who have collaborated, as partners, both for the financial part and for the supply of the most innovative systems and solutions. The sponsors have also favored the diffusion of the results reached by the team members on a national scale and not from time to time.

The project in fact involves a multidisciplinary team of about 50 students with the aim of creating a full-sized house that uses solar energy, and at the same time is green and smart, a sustainable and self-sufficient energy prototype. The team is made up of students, masters' students, doctoral students, and professors from the three Faculties of Architecture, Civil Engineering and Political Science, Sociology and Communication. The team's approach is based on the active involvement of the individual team members, through a holistic way of thinking.

There are four reasons why the Faculty Advisor Marco Casini took part in this competition. The first motivation concerns the University, or the possibility to bring Sapienza and its image in an international context and to deal with emerging foreign countries, such as the host country, the Emirates. The second reason is inherent in the work for students. The Solar Decathlon allowed them to relate to the world of work, to focus on a concrete project, in work that is not only theoretical, but that allows them to move from the idea of the project to its physical construction in detail. The third motivation concerns the possibility of doing research on various aspects, from strictly design and construction aspects, related to the house, to those concerning the sociological part and the behavior of people. It is a real laboratory where you can experiment. The interesting thing is that this experimentation is done together with the sponsors, building a relationship between university and industry which is what is missing in the Italian scenario, typical instead of the Anglo-Saxon and American culture, where there is a great union between companies and universities to do applied research. Finally, one reason is represented by the possibility of increasing student awareness of sustainability and allowing them to imagine, design, and implement measures to improve energy efficiency through renewable sources. The latter is perhaps the most important point given that it is oriented to the discussion of the criticalities that our planet is experiencing in terms of environmental pollution, depletion of non-reproducible resources and climate change.

The project, realized with BIM technology, intends to face the challenges of society proposing a new architectural approach that manages to make the use of alternative technologies coexist in a planning attentive to the environmental context, with a training path aimed at use of sustainable technologies and innovative construction solutions. *According to the revolutionary 4.0 Architecture model, the ReStart4Smart project of the Sapienza Team intends to apply and test the most advanced tools, materials and technologies available to the building industry today in order to create a sustainable home that is able to respond to the numerous needs of efficiency, comfort, safety and economy posed by the Architecture of the 21st century.*

The Smart Solar House was in fact an open laboratory for learning and training for sustainability. Intelligence and flexibility have guided the optimization of the functioning of the house so that it could vary adapting to the growth needs not only of the family unit, but also of the building fabric of the city. It was built, before being disassembled and then presented in Dubai, in the CEFME-CTP center in Pomezia (RM), which allowed the team to carry out the necessary assembly tests. The competition included a house of about 90 square meters, the Sapienza project fully respected these standards by building a completely wooden house of about 80 square meters, with a shady interior patio designed for the climate of the Middle East and which recalls both the Arab and the Roman culture.

The project was divided into four pillars, hence the name ReStart4Smart: a typological level (smart shape), a technical-constructive level (smart envelope), a technological level (smart systems), and a socio-cultural level (smart people).

Smart shape concerns the orientation of the building and the distribution of interior spaces to favor the exploitation of renewable energies, the reduction of energy requirements, as well as natural lighting and ventilation. The goal of this pillar and of the group of students working in this area is to design a Solar House that integrates perfectly with the climate in the Middle East. It also wants to propose a model of home that marries the Arab tradition characterized by a modular and replicable building typology able to adapt to different urban fabrics. Smart system that through the synergistic use of cutting-edge technologies and high-efficiency plant solutions aims at reducing primary energy consumption from non-renewable sources and maximizing internal comfort levels; the maximum integration between the various forms of solar energy suitable for the Middle East is promoted, the design of control systems and environmental monitoring and the reduction of water consumption by over 65%. Smart envelope wants to maximize flexibility and resilience by reducing production costs and energy requirements; the specific objectives of the work group are traceable in the ability to design a type of flexible and expandable building to demonstrate the great capacity of the wood capable of guaranteeing an optimal level of performance depending on costs. Smart people, who in a socio-cultural context aim at the socialization, that is domotics, of the individual through the training and direct involvement of users. The aim was to design intelligent buildings perfectly integrated in the Middle East context that are connected both with a global network and with users and able to share their functions with the city in order to be bases for the creation of the smart city of the future.

After the end of the competition, the project remained alive, field research activities continued with the partners and experimentation with new forms of energy efficiency alongside Enea. The project was then awarded at the Milan Triennale as one of the hundred Italian excellences in the construction sector '100 Italian Stories for Future Building' for the 'design and management' category.

Ultimately ReStart4Smart is not just a home, but a large network of activities that focus on the user in all its facets, from interaction with the smart home, to learning energy saving technologies, to inclusion in an increasingly green and smart company.

Conclusions

The contributions present in this chapter have been intended to highlight how the five dimensions of smartness are closely related to each other. Although the smart society is the great absentee from the point of view of literature, it, as highlighted, represents the central dimension of the discourse. In fact, society as a social system, as a set of ties and relationships, as economic, political, working, and housing cohesion, is what makes the smart dimension possible.

Smart politics has brought to attention how smartness cannot be considered political free, but how, on the contrary, the political dimension is contained in it. The main risk that has been highlighted lies in the danger of an excessive concentration of power in favor of the economic-financial and technological spheres. For this reason, it is essential that the discourse on smartness regains the dimension of political character, which is too often set aside.

It was then emphasized that smart people, through their peculiar characteristics, are the central social actors of this smart world. Intelligence is, in this sense, implemented through high levels of qualification, propensity to learn, open-mindedness and creativity.

The smart community, moreover, represents, as we have seen, the dimension that is formed when a certain type of relationality between the subjects is established through which it is possible to generate alliances with the aim of improving the quality of life, mediating between institutions and companies, and employing new technologies, applying them to inclusion.

Finally, we wanted to highlight how the smart home comprises all the previous dimensions, in other words, the house represents a microcosm in which the smart dimensions coexist in a systemic and dynamic way. Integration, inclusion, and governance are therefore the key words that link the reality of the subject in the community, of the community with politics, of politics with respect to society.

The originality of the solar house is found, therefore, in having managed to incorporate the smart facets described in this volume. The smart solar house has highlighted how the synergy between the various dimensions of the smartness is possible and materially achievable. In fact, the most critical aspect of smart technologies is often the proof of reality, while the solar house was designed, as mentioned, to respond to the personal, territorial, and social needs of the context in which it was set.

The work completed by the team, allowed by the funding allocated by the European Union, which, as is known, has always shown an interest in technology applied to social issues, has revealed how the dimension of smart politics enters into the realization of smart projects in the close union that binds it to the community and to people. The common goal for all dimensions – smart politics, smart community, and smart people – is therefore that of the inclusion and development of new technologies for the construction of a smart society that allows the smart home to be fully integrated, not only technological but also and above all social.

Notes

1 This chapter is the result of a survey designed and carried out by the above listed authors. However, for a more accurate attribution, we owe to Romina Gurashi the elaboration of par. 2 'A European vision: Horizon 2020', to Ilaria Iannuzzi the elaboration of par. 3 'The competition: Solar Decathlon Middle East 2018', and to Melissa Sessa the elaboration of par. 3 'The case study: ReStart4Smart'. All the authors have equally contributed to the writing of the introduction and conclusion.

2 With regard to funding, in H2020 there is a single funding rate for all beneficiaries and all activities: EU funding covers up to 100% of all eligible costs for research and innovation actions (i.e. for research at wide range), while funding generally covers 70% of the eligible costs for innovation actions (i.e. for research closest to the market. The share can reach 100% for non-profit organizations). In addition to the direct eligible costs (reimbursed based on the reporting) an additional fixed flat rate of 'indirect costs' (general expenses of the organization) corresponding to 25% of the direct eligible costs is reimbursed.

3 The H2020 program is the successor of the 7th Framework Program (FP7) implemented in the 2007–2013 period. Compared to FP7, H2020 has a greater financial allocation and provides for a considerable simplification in the rules for participation, as regards the structure of the loans, the management methods and the evaluation criteria (excellence, impact, quality, and effectiveness of implementation).

4 Ajman University of Science and Technology (United Arab Emirates); American University in Dubai (United Arab Emirates); American University of Ras AlKhaimah (United Arab Emirates); Dhofar University (Oman); Eindhoven University of Technology (Netherlands); Heriot-Watt University Dubai (United Arab Emirates); Gabriele D'Annunzio University of Chieti-Pescara, University of Pisa, Second University of Naples, University of Sassari (Italy); Islamic Science University of Malaysia, University of Technology – Malaysia (Malaysia); Ion Mincu University of Architecture and Urbanism, Technical University of Civil Engineering Bucharest, Birla Institute of Technology and Science, Pilani in Dubai, Polytechnic University of Bucharest (Romania-United Arab Emirates); National Chiao Tung University (Taiwan); National University of Sciences and Technology (Pakistan); New York University Abu Dhabi (United Arab Emirates); Qatar University (Qatar); Sapienza University of Rome (Italy); University of Wollongong (Australia) The Petroleum Institute; Zayed University (United Arab Emirates); University of Sharjah, University of Ferrara (United Arab Emirates-Italy); University of Bordeaux, Amity University, An-Najah National University, Arts & Métiers Paris Tech, Bordeaux

School of Architecture (France-United Arab Emirates-Palestine); Virginia Tech (United States); King Saud University (Kingdom of Saudi Arabia); The University of Jordan (Jordan); University of Belgrade (Serbia).

5 All information reported here was taken from the official website of the Italian project http://www.restart4smart.com/it/. In addition to the project and the activities carried out by the team, the site also features the complete press review, which has recorded progress and successes of the project step by step, and which boasts titles such as *Il Fatto Quotidiano, Repubblica, Il Corriere della Sera, Il Sole24ore, La Stampa*, etc.

Index